John Joseph Flather

Rope-Driving

A Treatise on the Transmission of Power by Means of Fibrous Ropes

John Joseph Flather

Rope-Driving

A Treatise on the Transmission of Power by Means of Fibrous Ropes

ISBN/EAN: 9783337139285

Printed in Europe, USA, Canada, Australia, Japan

Cover: Foto ©berggeist007 / pixelio.de

More available books at **www.hansebooks.com**

ROPE-DRIVING:

*A TREATISE ON THE TRANSMISSION
OF POWER BY MEANS OF
FIBROUS ROPES.*

BY

JOHN J. FLATHER, Ph.B., M.M.E.,
Professor of Mechanical Engineering, Purdue University.

FIRST EDITION.
FIRST THOUSAND.

NEW YORK:
JOHN WILEY & SONS.
LONDON: CHAPMAN & HALL, LIMITED.
1895.

PREFACE.

THE following treatise has been prepared to supply the existing need of a comprehensive manual of practical information concerning rope-driving and the principles upon which the practice rests.

In its preparation free use has been made of whatever literature could be found relating to the subject, and references for further investigation are given in foot-notes throughout the work.

Most of the data, however, have been collected by the writer, who desires to acknowledge his indebtedness to those who have assisted in the work by furnishing information and drawings. Especially to be mentioned are Mr. C. W. Hunt, the well-known authority upon this subject; and Mr. Spencer Miller, long identified with the practice of rope transmission.

J. J. FLATHER.

LAFAYETTE, IND., Oct. 1895.

CONTENTS.

CHAPTER I.
INTRODUCTION.. 1
    Leather belts and spur-gears—Early use of ropes—Advantages of ropes—Losses in various systems of transmission.

CHAPTER II.
MULTIPLE ROPE SYSTEM.. 10
    Rope-wells—Distribution of power to several floors—Influence of arc of contact—Rope-splicing.

CHAPTER III.
CONTINUOUS ROPE OR WOUND SYSTEM............................. 25
    Vertical transmission through several floors—Tension-carriages—Rope-tightener—Contraction of ropes—Transmission at an angle—Side lead—Multiple idlers—Dynamo driving—Jack-shafts—Coil-friction—Winder-pulleys—Application of ropes to water-wheels—Rochester plant.

CHAPTER IV.
LONG-DISTANCE TRANSMISSION.................................. 62
    Hirn's use of steel band—Wire ropes—Use of shafting—Limit of length for shafting—Draw-rods—Efficiency of power transmissions.

CHAPTER V.
FIBROUS ROPES.. 75
    Rawhide—Leather—Manilla—Cotton—Structure of cotton fibre—Strength of cotton ropes—Manilla fibre.

CHAPTER VI.
MANUFACTURE OF ROPES.. 85
    Size of yarns—Degree of twist—Lambeth rope—Lubrication of ropes—Stevedore rope—Effect of tar on ropes—Area of rope—Strength of manilla ropes—Factor of strength and wear—Working strain—Size of ropes.

CHAPTER VII.

WEAR OF ROPES.. 103
    Influence of pulley diameter—Internal wear—External wear—Harmonic vibration—Life of ropes—Weight of ropes.

CHAPTER VIII.

HORSE-POWER TRANSMITTED BY ROPES...................... 111
    Coefficient of friction—Difference in tensions—Influence of centrifugal force—Graphic representation—Relative cost of rope-driving.

CHAPTER IX.

DEFLECTION OF ROPES... 128
    The catenary—Approximate equations—Table of deflections—Method of laying out curve—Inclined transmissions—Tension-weight for given deflection.

CHAPTER X.

LOSSES IN ROPE-DRIVING.. 141
    Engine-friction—Effect of temperature—Friction of shafting—Power absorbed by shafting.

CHAPTER XI.

LOSSES IN ROPE-DRIVING—*Concluded*......................... 159
    Resistance due to bending—Wedging in the grooves—Effect of groove angle—Wood rims—Differential driving—Creep of ropes.

CHAPTER XII.

CONSTRUCTION OF ROPE PULLEYS 177
    Least diameter of pulley—Uniformity of pitch—Milled grooves—Cast grooves—Light pulleys—Wood-filled pulley rims—Rim sections—Proportions for groove—Idlers—Hubs—Double arms—Split pulleys—Diameter of bolts—Sections of arms—Stresses in pulley—Methods of joining arms and rim—Built-up rope-pulleys.

TABLES.

	PAGE
I. Limit of Length for Steel Shafting	70
II. Limit of Length for Wrought-iron Shafting	71
III. Strength of Cotton Transmission Ropes	81
IV. Strength of Manilla Transmission Ropes	95
V. Executed Rope Transmissions	96
VI. Greatest Revolutions per Minute for given Diameter of Rope	104
VII. Weight of Ropes	110
VIII. Values of $1-z$ for a Working Stress Equivalent to $200d^2$ pounds	117
IX. Angle embraced by Rope	117
X. Friction and Stress Moduli	118
XI. Horse-power Transmitted by Ropes	121
XII. Relative First Cost of Rope-driving	123
XIII. Relative Wear and Cost of Rope per Horse-power Transmitted	125
XIV. Deflection of Rope	134
XV. Friction Per Cent under Varying Loads	143
XVI. Power-absorbed by Friction in Line-shaft	152
XVII. Power absorbed by Friction in Jack or Head Shafts	154
XVIII. Power absorbed by Lead-shaft carrying High-speed Ropes	156
XIX. Angle of Grooves for Equal Adhesion	169
XX. Values of $d^{1.1}$ and $\sqrt[4]{V}$	180
XXI. Least Diameter of Pulley for given Diameter and Speed of Manilla Rope	180
XXII. Rope-pulleys for General Work	181

ROPE-DRIVING.

CHAPTER I.

ALTHOUGH toothed gearing and belts are the most familiar mediums for the transmission of power in mills and factories, systems of rope transmission for this purpose have been in use for many years, but it is only within the last ten years that they have given promise of being generally recognized in this country as a convenient and efficient means of accomplishing the ends for which they were designed. Until about fifty years ago it was generally thought by engineers that cotton- and woollen-mills, and all others requiring a considerable amount of power, could not be run effectively without large and ponderous lines of upright and horizontal shafts of either cast or wrought iron and heavy trains of gear wheels.

When large leather belts began to be introduced as a substitute for gears it was thought to be an experiment of very doubtful result, if not altogether impracticable;* but when high speeds and lighter shafting were used in conjunction with the wide belts, the marked success which attended their general adoption in America during the next twenty years attracted considerable attention in England. It was not, however, until about thirty years ago that the

---

* *Journal Franklin Inst.*, 1837.

American system began to replace the old-fashioned gearing. The belt system made very slow progress in England, however, and before it had been at all extensively adopted a newer method was introduced and quickly came into prominence, making such rapid progress as to almost entirely supplant the old wheels.

The use of hemp rope for power transmission had been revived about 1860 by the Messrs. Combe & Barbour of Belfast, who introduced it successfully into small mills in the north of Ireland. This was followed by its speedy adoption in the jute-mills of Dundee, and subsequently in the cotton-mills of England.

Previous to this, fibrous ropes had been in use for transmission purposes, but their application had been limited.

At the Colonial Rope Factory at Great Grimbsy, in Lincolnshire, from 1830 to 1837, ropes had been employed for taking off the power from the engine and communicating to the first motion shaft. The plan was very commonly adopted in connection with rope-works, where the driving rope employed was known as the fly-rope for working the equalizer from end to end of the ropewalk. The machinery in connection with the flax-spinning mills at the same place was also driven by means of rope gearing.*

Ropes were also used in this country many years before their more general adoption in England which followed the movement introduced by Messrs. Combe & Barbour. In a recent communication, Mr. Samuel Webber states that he remembers the occasional use of rope-driving for temporary purposes "back in the forties." In one case he mentions, power was carried from a small engine outside through a window into a mill to grind cards before the wheels and main belts were ready.

The use of ropes, however, was not common, and it has

---

* Mr. W. Smith, Proc. I. M. E. 1876, p. 393.

only been in recent years that rope-driving has come into prominence as a factor in power transmission.

By this system, according to one of its earlier promoters,* Mr. Jas. Durie, large powers were transmitted "by means of round ropes working on grooved wheels, which in some parts of this country [England] have been largely substituted for toothed gearing. In this mode of driving, the fly-wheel of the engine is made considerably broader than the fly-wheel of an engine having cogs on its circumference; and, instead of cogs, a number of parallel grooves for the ropes are turned out, the number and size of which are regulated by the power to be taken off the fly-wheel. The power which each of the ropes will transmit depends upon their size and the velocity of the periphery of the fly-wheel."

As rope-driving has, until recent years, been a matter largely of experiment, the results which have been obtained from its use have not always been of uniform excellence, mainly, however, because designers have failed to properly recognize the requirements.

As the conditions under which the systems were installed have been so varied, it is not surprising to find many cases where the ropes have been rapidly worn out and replaced by leather belting, or other methods of transmission; but where rope-driving has been tried and has failed, investigation will invariably show the absence of suitable conditions or a disregard of correct principles of design or construction. In many applications too great a strain is put upon the rope, and the stretch and wear are rapid; in addition to this, the pulleys are often of unsuitable size, and the rope is unnecessarily weakened through the action of the fibres upon one another. Both of these causes are a constant source of annoyance in a rope-drive which has been poorly designed or constructed. A case in point can

---

* Proc. I. M. E. 1876.

be shown where a rope would stretch to such an extent that it had to be taken up every few days when new, and later every two or three weeks—a rope under these conditions lasting less than four months. Investigation showed that a tension-weight of 600 pounds had been placed upon the rope, which was of ¾ inch diameter. A weight of 50 pounds was substituted, and the rope has since been running satisfactorily for three years, and has only been taken up once in that time.

Cases where ropes have suddenly broken are few in number, the risk in this respect being reduced to a minimum by the fact that any defect in a rope, arising either from wear or other cause, will show itself long before the point of danger is reached.

The ropes by which the power is transmitted consist of an elastic substance, and their lightness, elasticity, and comparative slackness between the pulleys are highly conducive to their taking up any irregularity that may occur in the motive power.

Their quiet working and convenience in application, much more so than wide belts, are also features which caused ropes to be looked upon with great favor. Another reason for its rapid progress in England, which was considered a great advantage by the millowners, was the adoption of the multiple rope, now known as the English system, in which a number of single ropes are spliced and run side by side.

The entire freedom from any risk of a breakdown or stoppage of the works which might occur with gearing was an important factor in replacing the latter by the newer system; the working stress in the ropes being but a fraction of their breaking strength, any signs of weakness in an individual rope would allow it to be removed, and the engine run with the remaining ropes until a convenient opportunity offered for the replacement of the weak member.

ROPE-DRIVING.

One of the great advantages of rope-driving over gearing lies in the steady motion produced, but this has been attributed more to an accidental combination of heavy flywheel and high velocity than to any inherent advantage in the system itself.

In spur gears diameters of 20 to 24 feet were usual in large engines, while for ropes the pulleys are 25 to 35 feet in diameter; and, whereas the gears weighed from 20 to 25 tons, we find rope pulleys used for the same work weighing from 60 to 75 tons.*

When we consider the speed at which these heavy wheels are run—from 3000 to 6000 feet per minute—it is not surprising that uniform rotation is obtained; and whether it be that the energy stored in the moving mass prevents fluctuation, or whether the elasticity and other properties of the rope perform the same office, or—which is more reasonably the case—all of these factors act together, the truth remains that steadier running and a greater output are now obtained with rope-driving than was formerly the case with toothed gearing.

The general experience is not altogether in favor of ropes, for, while the advantages of smooth running and easy handling are conceded, it is also acknowledged that the extra weight and greater width of pulleys increase the journal friction over that found with toothed gearing, and that otherwise a greater loss of power occurs, the causes of which will be discussed subsequently.

---

* The Walker Manufacturing Co. of Cleveland recently made four large rope pulleys for the Broadway Cable Railway Company, New York, which weighed 104 tons each. These were 32 feet in diameter and were grooved for thirty-four 2-inch ropes.

Another example of heavy rope wheel is given in London *Engineer*, January 11, 1884. This wheel was made by Hick, Hargraves & Co , Bolton, England, for a cotton mill in India, and is 30 feet in diameter, 15 feet face, grooved for 60 ropes to transmit 4000 horsepower. Its weight is 140 tons.

It is difficult, however, to determine the relation between the power absorbed by ropes and gears, for in nearly all cases where rope-driving has been substituted for gears, other changes have been made at the same time, or the engines were, after the alteration, driven at an increased speed, so that there has been little opportunity for comparison. There is, however, a generally-accepted opinion among engineers that the loss in rope-transmission is from 5 to 10 per cent greater than with gearing.

Mr. A. G. Brown* states that in the older cotton-mills of England, where the main drives are by gearing, and belting is used for the intermediate and machine driving, the friction of the engine, shafting, gearing, and belting averages about 20 per cent of the whole power, the engine developing at full loads from 500 to 800 I. H. P.

These engines were compound condensing, and consisted in each case of an overhead beam engine which had been converted into a compound by the addition of a high-pressure cylinder between the crank and beam centre.

In the newer mills, for doing practically the same work, and where the main and some of the intermediate drives are by ropes, the friction of the engine, ropes, shafting, and belting averages 23 to 25 per cent. of the whole power, the engines developing from 800 to 1500 I. H. P. It is fair to assume that the newer engines have at least no greater percentage of friction than the older ones, except that due to an increased journal friction attributable to the larger journals and greater weight of the rope-pulleys; also the friction of the intermediate shafting and small belts should be the same in each case; therefore it is reasonable to conclude that the increase of power absorbed by the use of rope-driving is chargeable to the system itself. The writer has assumed in similar cases that the loss at the engine in

---

* *American Machinist*, July 21, 1888.

rope driving is about 10 per cent of the indicated horse-power, that an additional 10 per cent is absorbed by the mill-shafting, and that from 5 to 8 per cent may be attributed to losses in the rope itself due to resistance to bending, wedging in the grooves, differential driving effect, and creep, all of which affect the loss to a greater or less extent. As compared with the above, the friction of shafting and engines in American cotton-mills, where belting is used exclusively, indicates that the percentage of loss where large belts are employed is probably a trifle less than that obtained with ropes in English mills ; but the conditions of practice are so varied that it is difficult to compare the two systems from published results of tests. As shown on page 155 for similar installations, the loss absorbed in shaft-friction will not materially differ in the two systems; for large transmissions the engine friction should be less with ropes, but the losses in the ropes themselves due to slip, differential driving, bending, creep, and other causes, may, without special precautions, exceed to a small degree the losses due to the belt.

Mr. J. T. Henthorn, in a paper read before the American Society of Mechanical Engineers, states that the friction of the shafting and engine in a print-mill should not exceed 19 per cent of the full power; but out of fifty-five examples of a miscellaneous character which he has tabulated only seven cases are below 20 per cent, 20 vary from 20 to 25 per cent, fifteen from 25 to 30 per cent, eleven from 30 to 35 per cent, and two above 35 per cent. We note that the greater number lies between 20 and 25 per cent; allowing a variation of 5 per cent each side of these limits we shall obtain values from which a fair average may be determined. This will include those cases under 20 per cent, but not those over 30, from which we find the mean loss to be 23.9 per cent of the total power.

Mr. Barrus, speaking of this subject, quotes eight cases,

the data of which were obtained from tests made by himself in various New England cotton-mills, in which the minimum percentage was 18 and the maximum 25.7, the total average being 22 per cent.

Mr. Samuel Webber states that 16 per cent of the total power of a mill is sufficient to overcome the friction of shafting and engine—10 per cent for the shafting and 6 per cent for the engine. But in this estimate Mr. Webber does not include the loss due to the belts running upon loose pulleys, which he does not consider to be part of the shafting, as they are not so running while the machinery is in operation ; and when it is not, they may be thrown off as well as not, except for convenience.

He further estimates, both from his own experience and the observations of others, that the power consumed by the machine belts on the loose pulleys in a large cotton-mill is about 8 per cent of the whole. This 8 per cent added to the 16 per cent loss due to shafting and engine will give 24 per cent of the total power—a result which agrees closely with the average values given above.[*]

Considering the greater loss which occurs in the use of ropes and belts for main drives, the recent revival of gearing for this purpose in England has much in its favor. Of this Mr. Geo. Richards states:[†] "The advantages of rope transmission for main drives in large plants would not be as apparent if compared with modern gearing. The kind of gear used for this purpose to-day is not the rough cast gear used formerly, whose uneven motion produced a rumbling which could be heard a mile or two from the mill. The present gears are often machine-cut, made to bear equally on each tooth, and with a contact

---

[*] See "Dynamometers and the Measurement of Power," John Wiley & Sons, New York.

[†] Richards's "Mechanical Progress," 1891.

across the whole face of the gear, causing little more noise than the ropes.

With greater speed and stronger and heavier gears steady running is insured, and by using properly-proportioned machine-cut teeth comparatively little noise results. The saving of space is also an important factor advanced in favor of gearing." At the present time steel gears 30 feet in diameter are being made, with all the teeth cut, for transmitting the powers of mill-engines in the Oldham district, while in this country machine-cut gears from 30 to 50 feet in diameter are in use.

However, although such gearing may be very superior to the former slow-running cast gears, it is questionable whether it can ever produce the same steadiness of running which is so largely a distinguishing feature of rope-transmission. Any shock or sudden fluctuation of load must necessarily be transmitted through the gear teeth, whereas with rope transmission such shock is partially absorbed by the more or less elastic ropes and subsequently given out by their recoil.

For this reason when uniformity of speed is desired ropes are generally to be preferred to gears, even when the latter are working under their most advantageous conditions.

CHAPTER II.

WHERE ropes have been used to replace gearing in the English mills the plan adopted has been to put in a new grooved fly-wheel, or to place grooved segments upon the existing fly-wheel, when the speed could be increased sufficiently to allow of a limited number of ropes being employed, and the width of the wheel-pit was also sufficient for the purpose; but if this plan could not be adopted grooved pulleys were put on the intermediate shaft, and the ropes carried to the different stories of the mill. It has sometimes been necessary to put in a countershaft, so as to gain speed and obtain a sufficient distance between the centres of the shafts on which the pulleys are placed.

Where rope-driving has been installed in new factories special provision has been made for the ropes, and we find in such cases rope-wells or chambers built in, suitably fitted with platforms and staircases to give access to pulleys and bearings on the various shafts, as shown in Figs. 1 and 2.

The majority of drives are arranged so that the ropes are horizontal or inclined rather than vertical, and with the driving or tight side of the rope on the lower side of the pulleys; then, when transmitting power, the two sides approach each other, and the arc of contact is increased.

An additional advantage is that obtained by the weight of the rope acting on both pulleys, thus allowing a low initial tension to be maintained. This does not hold for short distances between centres, as under such conditions the weight of the rope adds little to the total tension; on the other hand, where the distance between the pulleys in

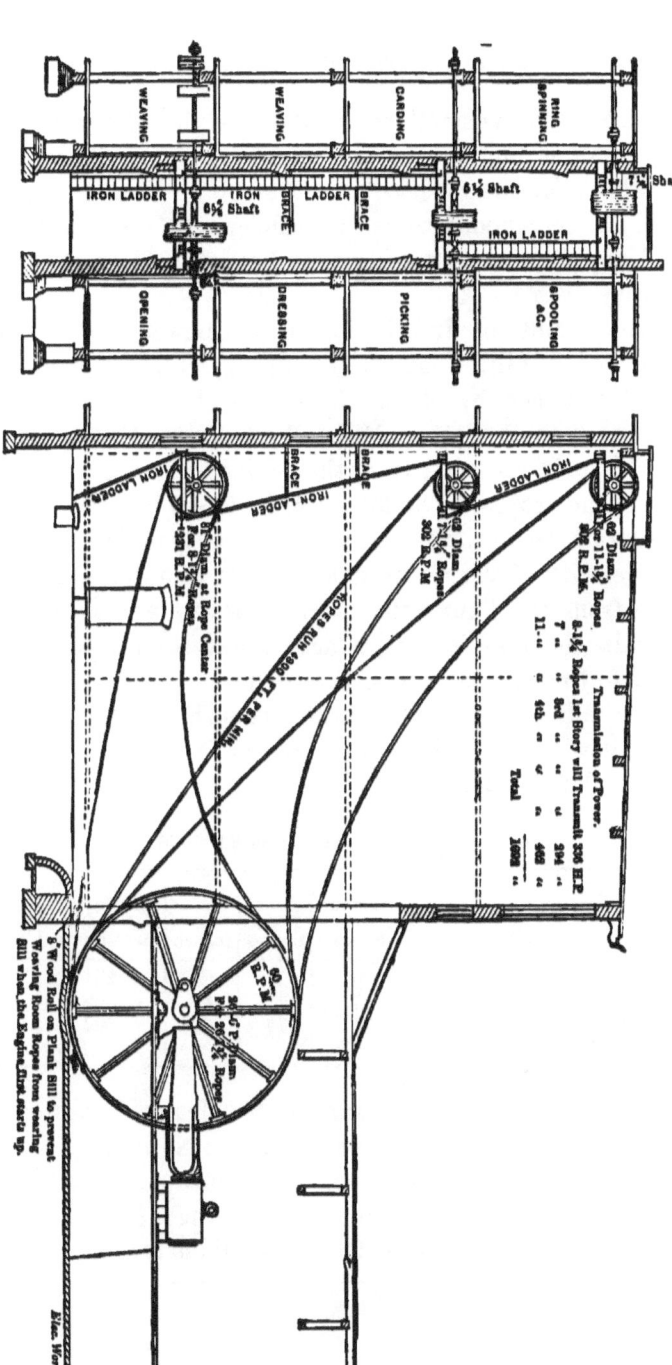

Fig. 1.—Method of Distributing Power to Several Floors, Lanett Cotton Mills, Georgia.

vertical drives is small, the relative weight of the rope being small as compared with its tension, there will be little tendency for the rope to leave the bottom sheave.

With these conditions the efficiency of vertical drives will approach that of a horizontal or inclined arrangement of ropes.

The manner of distribution of the power to the several floors of a mill is shown in Fig. 1, which represents a plant designed by Messrs. Lockwood & Greene of Boston for the Lanett Cotton Mills, West Point, Ga., in which 1100 h. p. is delivered from the engine by means of twenty-six $1\frac{3}{4}$-inch ropes. The fly-wheel is 26 feet in diameter, and makes 60 revolutions per minute, corresponding to which the velocity of the rim is 4900 feet per minute. As will be seen from the figure, the driving-sheaves are placed in a well in the middle of the factory, and the line-shafts extend to the right and left, as shown.

There is no line-shaft on the second floor, as the various machines may be driven from below.

The distribution is as follows:

ROPE-DRIVES IN LANETT COTTON MILLS.

	Number of Ropes.	Dia. of Rope. Inches	Dia. of Pulley. Inches.	Revolutions of Pulley per minute.	Horse-power.
1st floor.........	8	$1\frac{3}{4}$	81	231	336
3d floor.........	7	$1\frac{3}{4}$	62	302	294
4th floor.........	11	$1\frac{3}{4}$	62	302	462

A similar plant, also designed by Messrs. Lockwood & Greene, has been recently erected at the Naumkeag Cotton Mills, Salem, Mass., in which 1800 horse-power is distributed to five floors by means of forty-one $1\frac{3}{4}$-inch manilla ropes.* The fly-wheel is 26 feet in diameter, by about $9\frac{1}{2}$

---

* *Power*, March 1895.

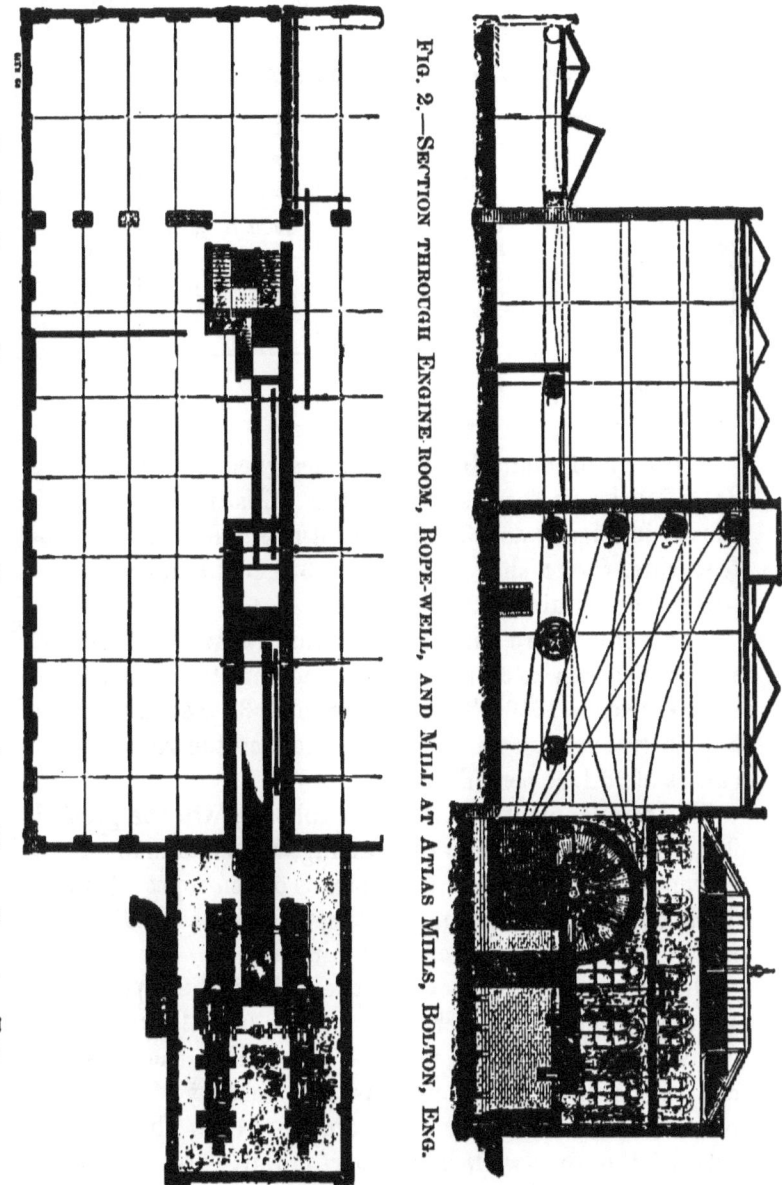

Fig. 2.—Section through Engine-room, Rope-well, and Mill at Atlas Mills, Bolton, Eng.

Fig. 3.—Plan of Engine-room and Rope-well at Atlas Mills, Bolton, Eng.

feet face, and weighs 150,000 pounds. The velocity of the ropes is the same as in the previous instance, namely, 4900 feet per minute. The distribution is as follows:

	Number of Ropes.	Dia of Rope.	Speed of Rope.	Horse-power.
1st floor.........	13	1¾ inches.	4900 ft p. m.	585
2d "	14	"	"	630
3d "	4	"	"	180
4th "	5	"	"	225
5th "	5	"	"	225
Total......	41			1845

A system of main driving-gear designed and erected by John Musgrave & Sons at the Atlas Mills, Bolton, Eng., is shown in Figs. 2 and 3. This mill is 300 feet long by 135 feet wide, with a shed on one side 325 feet long by 45 feet wide, and contains 84,000 spindles.

The engines are tandem compound, 24 and 46 by 6 ft. stroke, and run at 50 revolutions per minute. The average horse-power is 1050. The rope-wheel is 32 feet in diameter and is grooved for thirty-two 1¾-inch cotton ropes, which run at 5026 feet per minute.

The arrangement of shaft is as follows: On the ground-floor there are five lines of shafting, the main shaft being driven from the rope-drum by means of ten ropes 1¾ inches diameter running on a pulley 9 feet 4¾ inches in diameter, which runs 170 revolutions per minute.

On the main shaft, close to the wall of the rope-well, is a pulley 6 feet diameter, grooved for four 1¾-inch ropes, which drives, through a similar-sized pulley, the line-shaft on the right. These ropes run at a velocity of 3205 feet per minute; also on the main shaft, but on the opposite side of the rope-well, is another pulley, 8 feet diameter, grooved for six 1¾-inch ropes, driving on to a pulley 64¼

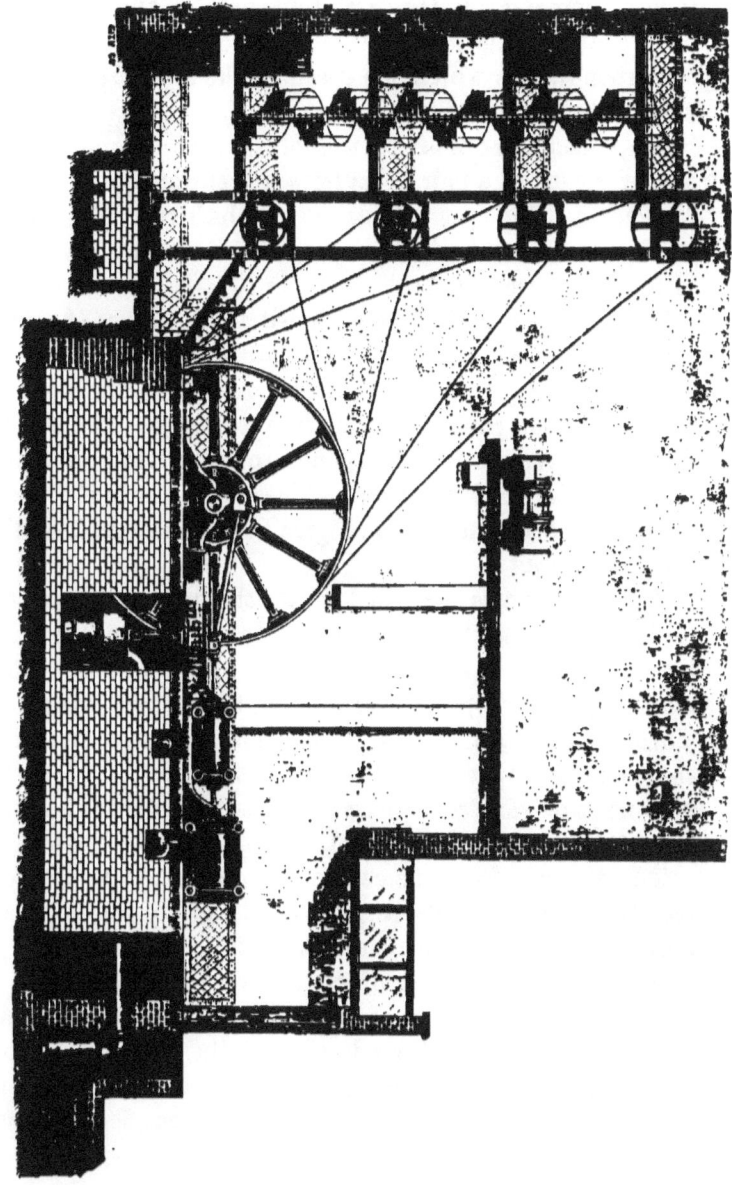

Fig. 4.—Section through Engine-room at the Thread Mills of the Nevsky Cotton-Spinning Co., St. Petersburg.

inches diameter on the first line-shaft to the left of the main shaft. This shaft runs 250 revolutions per minute, corresponding to which the speed of the ropes driving it is 4274 feet per minute.

The second line-shaft on the left is driven from the first one by means of a pair of pulleys 4 feet diameter, grooved for five 1¾-inch ropes. This second shaft runs 250 revolutions per minute, and the ropes driving it have a velocity of 3140 feet per minute.

The second line-shaft, mentioned above, drives the line-shaft in the shed by means of a pair of pulleys 4 feet diameter, grooved for four 1¼-inch ropes, driving the counter-shaft shown on plan, on which is a pulley 3 feet 4 inches diameter, grooved for five 1¼-inch ropes, driving on a 3-ft. pulley on the shed line-shaft. These ropes have a velocity of 2640 feet per minute and give to the shaft 278 revolutions per minute.

All of the shafts described are on the ground-floor of the mill. Of the shafts above this, on the next two floors, the line-shafts each have pulleys 6 feet in diameter, grooved for seven 1¾-inch ropes, driven from the main rope-drum. These shafts run at 266 revolutions per minute. The shaft on the upper floor also runs 266 revolutions per minute and is driven from the main rope-drum through a pulley 6 feet in diameter grooved for eight 1¾-inch ropes.

The distance from the centre of upper shaft to the centre of crank-shaft is 89 feet, and the length of each rope required for this drive is about 250 feet.

Another arrangement of rope-drive for cotton-mills is shown in Fig. 4, which represents a section through the engine-room at the thread-mills of the Nevsky Cotton-Spinning Co., St. Petersburg.* In this drive there is no rope-chamber, as the whole of the rope gear is contained

---

* John Musgrave & Sons, engineers.

in the engine-room situated in the centre of the mill, which is 680 feet long and 90 feet wide.

There is a short staircase from the engine-room floor to the first landing, and the landings above this are reached by an ornamental spiral stairway as shown.

There are two shafts, one from each side of the engine-room; these are driven by a pair of right- and left-hand tandem compound engines, 30 and 52 by 6 ft. stroke, running at 50 revolutions per minute.

The average power developed by each engine is 1100 horse-power.

The rope-drums are each 30 feet in diameter and weigh 62 tons; these are grooved for twenty-eight 1¾ cotton ropes, which run at 4700 feet per minute.

The first-and second-floor shafts make 300 revolutions per minute, and are each driven by nine 1¾-inch ropes from the rope-drum running over pulleys 5 feet in diameter on the shafts.

The shafts on the third and fourth floors run 200 revolutions per minute, and are driven by five ropes, each 1¾ inches diameter, running over 7 feet 6 inch pulleys on the shafts.

In each of the two upper rooms there is a second line-shaft, driven from the main line-shaft on each floor by means of 54-inch pulleys, grooved for four 1½-inch ropes, which have a velocity of 2826 feet per minute. The distance from the centre of the upper shaft to the centre of the crank-shaft of engine is 56 feet 6 inches. This is a short drive for a mill of this size; in fact, all of the drives are short, the lower one especially so, being only 80 feet between centres, the peculiar arrangement of the engine-room not admitting of a greater length; but the plant is said to work extremely well.

In these examples the multiple-rope system is used, each wind consisting of a separate rope stretched around the fly-wheel and its individual shaft-pulley, then spliced. The degree of tightness will depend upon the material of the rope and the amount of tension in the slack part necessary for adhesion. In the majority of cases the initial tension is very small compared with the strength of the rope—especially so where the horizontal distance between driving and driven pulleys is great, as, under such conditions, the tension in the slack side due to the weight of rope in the hanging catenary is often sufficient to prevent slipping; a slack upper rope in horizontal or inclined drives will also increase the arc of contact, thereby increasing the grip of the rope.

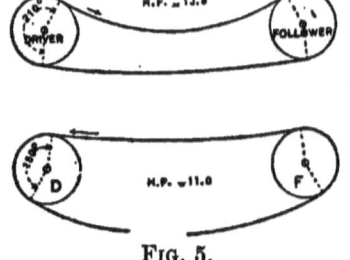

Fig. 5.

This is shown in Fig. 5, which represents two pulleys of equal diameter, arranged, as in the upper figure, to drive with the slack side uppermost, and, in the lower figure, with the slack side below.

The difference in the arc of contact, as shown in the figures, is 60 degrees, which would under similar conditions, with a velocity of 4000 feet per minute, produce a difference of over 25 per cent in the amount of horse-power transmitted by the two ropes under the usual working tension.

New cotton ropes are often stretched as taut as possible on account of their extensibility, as they will soon become slack enough for good working, and may even have to be respliced before becoming permanently set.

It is the practice of some engineers to strain both manilla and cotton ropes as much as possible and unite the ends with a temporary short splice when first put over the pulleys; after running a few days a permanent stretch is given to the rope, which is then respliced with a long splice, the strain on the rope being very much reduced in this latter case.

The splice in a transmission rope is not only the weakest part of the rope, but is the first to fail when the rope is worn out. If the joint is not strong the rope will fail by breakage or pulling out of the splice, the projecting parts will wear on the pulleys, and the rope fail from the cutting off of the threads. Formerly much trouble was experienced in this way on account of improper splicing. One form of joint, according to Cromwell,* was made by pressing the ropes firmly together and winding about with stout small rope. The spliced part is taken as long as possible in order to bend properly over the pulleys and give the required strength. As this form of joint made the rope larger in diameter at the splice, the effect produced was to run faster when passing over the driving-sheave and slower over the follower; the resulting motion was very irregular, and the wear at the splice rapidly destroyed the rope.

A very simple splice is sometimes used with rope-driving formed by opening out the ends of the rope for 12 or 15 inches and tying together the individual rope-yarns one by one, allowing the ends to lie straight, and serving the whole with spun yarn.

Similar joints wrapped with raw-hide belt-lacing give a very smooth splice which lasts well.

Some engineers favor a short splice, in that it is easily made and holds well, and offers a lesser length of enlarged portion for surface contact with the pulley.

If properly made, however, there need be no enlarged portion, and since a long splice is stronger we find such joints preferred in most cases.

There are several kinds of long splices varying in length from 60 to 80 diameters of rope, but the one which seems

---

* J. H. Cromwell, " Belts and Pulleys," John Wiley & Sons.

to give the best results in practice is the "English splice," directions for which are given in various trade publications.

The successive operations for splicing a 1¾-inch rope by this method are as follows: *

1. Tie a piece of twine, 9 and 10, Fig. 6, around the rope to be spliced about six feet from each end. Then unlay the strands of each end back to the twine.

2. Butt the ropes together and twist each corresponding pair of strands loosely, to keep them from being tangled, as shown at (*a*), Fig. 6.

3. The twine 10 is now cut, and the strand 8 unlaid and strand 7 carefully laid in its place for a distance of four and a half feet from the junction.

4. The strand 6 is next unlaid about one and a half feet and strand 5 laid in its place.

5. The ends of the cores are now cut off so they just meet.

6. Unlay strand 1 four and a half feet, laying strand 2 in its place.

7. Unlay strand 3 one and a half feet, laying in strand 4.

8. Cut all the strands off to a length of about twenty inches, for convenience in manipulation. The rope now assumes the form shown in (*b*), with the meeting-points of the strands three feet apart.

Each pair of strands is now successively subjected to the following operations:

9. From the point of meeting of the strands 8 and 7 unlay each one three turns; split both the strand 8 and the strand 7 in halves, as far back as they are now unlaid, and the end of each half strand "whipped" with a small piece of twine.

10. The half of the strand 7 is now laid in three turns, and the half of 8 also laid in three turns. The half strands

---

* From "Manilla Rope," C. W. Hunt Co., New York.

ROPE DRIVING.

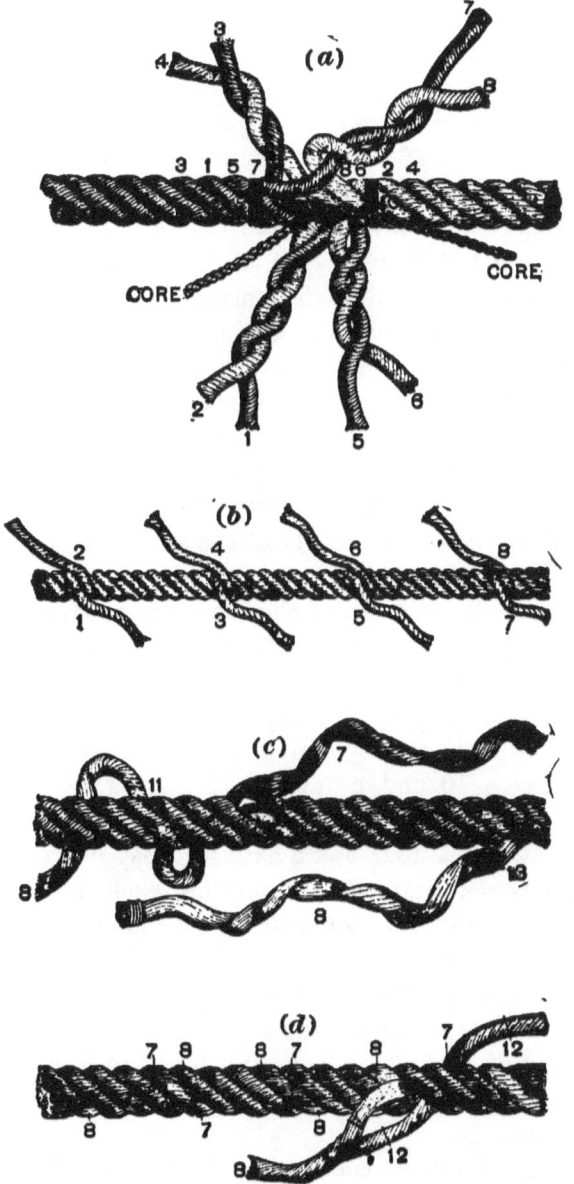

FIG. 6.—SPLICE FOR 1¾-INCH 4-STRAND ROPE.

now meet and are tied in a simple knot 11, (c), making the rope at this point its original size.

11. The rope is now opened with a marlinspike, and the half strand of 7 worked around the half strand of 8 by passing the end of the half strand through the rope, as shown, drawn taut, and again worked around this half strand until it reaches the half strand 13 that was not laid in. This half strand 13 is now split, and the half strand 7 drawn through the opening thus made, and then tucked under the two adjacent strands, as shown in (d).

12. The other half of the strand 8 is now wound around the other half strand 7 in the same way. After each pair of strands has been treated in this manner, the ends are cut off at 12, leaving them about four inches long. After a few days' wear they will draw into the body of the rope or wear off, so that the locality of the splice can scarcely be detected.

For a three-strand rope of the same size the foregoing method is slightly modified. After tying the twine 9 and 10 around the rope about 6 feet from each end, unlay the strands back to the twine, bring the butts together, and, as in Fig. 7, twist the corresponding strands loosely together. Now cut twine 10, and unlay strand 8 for a distance of four and a half feet from the junction, and lay in strand 7. Unlay strand 1 four and a half feet, lay in strand 2, and cut all the strands off to a length of about 20 inches, as before explained for convenience in handling. The splice now assumes an appearance similar to (b) with the exception that there are only three meeting-points of the strands, and these are $4\frac{1}{2}$ feet apart.

Each pair of strands is now subjected to the series of operations described for the 4-strand splice in steps 9 to 12 inclusive.

In splicing a Lambeth cotton rope the operation is modified to a still greater extent.

ROPE-DRIVING. 22a

Although considered as a troublesome rope to splice, the following instructions,* if carefully followed, will enable one to make an excellent joint without difficulty.

1. Tie a piece of twine 9 and 10 around the rope to be spliced about 6 feet from each end. Then unlay two strands together of each end back to the twine. Butt the

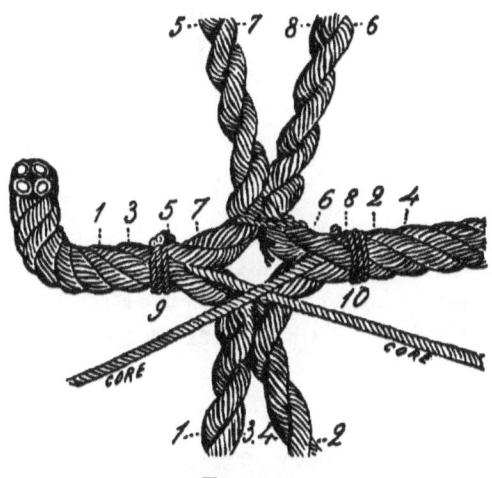

Fig. A.

ropes together and tie each set of strands temporarily, as shown at Fig. A.

2. The twine 10 is now cut, and the strands 6 and 8 unlaid together, and the strands 5 and 7 carefully laid in their places together for a distance of 18 inches. Then unlay strands 6 and 8, also 5 and 7, and tie strands 6 and 5 together temporarily. Next unlay strand 7 and lay in strand 8 in its place for a distance of 3 feet from strands 6

---

* Prepared for this work by the Manufacturers' Engineering Co., Boston, Mass.

and 5. Then tie strands 8 and 7 temporarily. Next cut off the ends of the core so that they will butt together. Strands 1, 2, 3, and 4 are next laid in the same manner as

Fig. B.

strands 5, 6, 7, and 8, but in the opposite direction. Care must be taken to keep the turns in the strands, or otherwise they will be soft and bulky. Next cut off all the strands

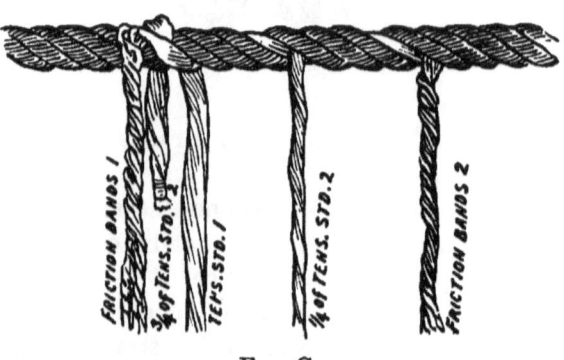

Fig. C.

to a length of about 24 inches, for convenience in handling. At this point the splice should be as shown in Fig. B.

The tension strand of a Lambeth cotton rope is the soft white yarn running through the centre of the strand, and is called the tension strand through its having to bear the strain put upon the rope in the transmission of power.

The friction bands of a Lambeth rope are the twisted outside yarns which are tubed around the tension strands to protect them from wear and contact in the grooves of the pulley.

3. Take strand 2 and unlay it two turns and remove the ten friction bands, then lay in tension strand 2 back again

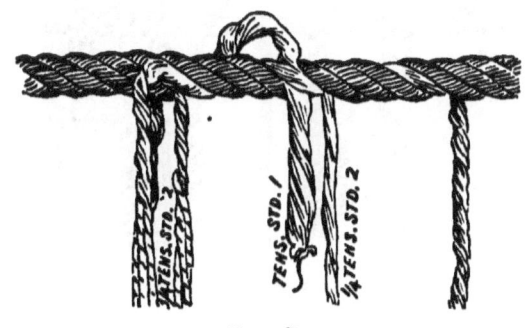

Fig. D.

one turn, split out ¼ of tension strand 2 and lay in the remaining ¾ of tension strand 2 for one turn. This will bring it to its former position. Remove the ten friction bands from strand 1, and tie tension strand 1 and ¾ of tension strand 2 in a simple knot. At this point of the knot the rope will be its original diameter, as shown in Fig. C.

4. Divide the friction bands removed from strand 1 in two parts, and take ¾ of tension strand 2, put it between the two parts and over tension strand 1 and through the

centre of the rope with the marlinspike. Next take tension strand 1 and work it around the $\frac{3}{4}$ of tension-strand 2 in the manner as shown at Fig. D.

5. Draw it taut and continue to work it around $\frac{3}{4}$ of tension strand 2 until it reaches the $\frac{1}{4}$ of tension strand 2; at this point $\frac{1}{4}$ of tension strand 1 must be removed, and continue to work $\frac{3}{4}$ of tension strand 1 around tension strand 2 until it reaches the friction bands removed from strand 2; divide these friction bands in two parts, and take $\frac{3}{4}$ of tension strand 1, put it between the two parts and

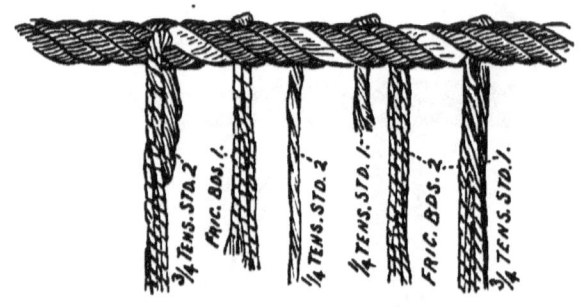

Fig. E.

over tension strand 2, and through the centre of the rope with the marlinspike. Next take the quarter of tension strands 1 and 2 and pass them through the centre of the rope on opposite sides with the marlinspike. Then half of the friction bands should be passed through the centre of the rope at each end with the spike. At this point the splice is complete, with the exception of cutting off the ends, and should be as shown at Fig. E.

6. The strands 3, 4, 5, 6, 7, and 8 should be worked in the same manner as 1 and 2.

Instead of using the ordinary marlinspike it will be found very convenient to drill out the body as shown in

Fig. 8, leaving only a thin shell four or five diameters deep.

By inserting the end of a strand in the bore of the mar-

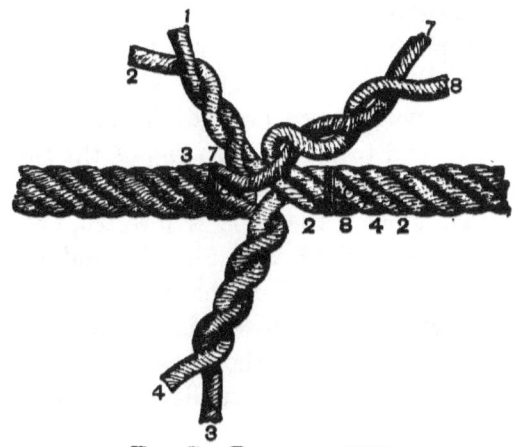

FIG. 7.—ROPE-SPLICING.

linspike the latter, with the strand, may be passed through and around the other strands as desired with much less trouble than ordinarily attends the operation.

FIG. 8.—IMPROVED FORM OF MARLINSPIKE.

For small braided ropes which cannot be spliced, a very convenient method of joining the ends is by the use of copper ferrules as shown in Fig. 9, which repre-

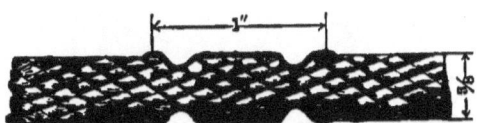

FIG. 9.—COUPLING FOR BRAIDED ROPE.

sents a form of joint devised by Mr. B. Frank Barnes, Rockford, Ill.

Samples of this splice were furnished by Mr. Jos. Burnett, which showed an efficiency of about 85 per cent. Thus in two samples the average breaking strength of the rope was 380 and 375 pounds respectively, while the splice pulled out under a strain of 320 pounds.

In this case the rope was a $\frac{3}{8}$-inch braided cotton cord which had been in use about three years. The coupling consists of a piece of copper tubing 1 inch long, into one end of which the rope is inserted about half way. A groove is then compressed around the tube and rope, by means of a special tool; the open end of the tube is then filled with sealing-wax and heated until the wax boils, then the other end of the rope is inserted, and the tube compressed. The melted wax fills the end of the rope, making a solid joint between the shoulders. With the large pulleys adopted (22 to 48 inches in diameter) no trouble is experienced, and the ropes last from two to three years, but the copper ferrules are changed about every four months.

Wood pulleys are used, and the grooves are filled with leather. Some of these ropes run as high as 5300 feet per minute. All the ropes are operated on the American or continuous wind system.

CHAPTER III.

A GOOD example of this system of rope transmission is shown in Fig. 10, which represents a plant designed by Mr. T. Spencer Miller, and erected for the Western Electric Company, New York, by the Link Belt Machinery Co. In this case vertical ropes are used, which are arranged to transmit the power of two 200-h. p. Russell engines, cylinders 18 by 27 inches, making 125 revolutions per minute; fly-wheels 10 feet diameter, each turned with eight grooves for 1⅛-inch rope. The ropes are of rawhide and wound continuously around the pulleys. As the rope leaves the fly-wheel at the left-hand side it runs over an idler, and from thence to a tension-pulley, or tightener, which is suspended in such a manner as to be drawn back by the weights, as shown. The arms which support the tighteners are hung from rollers, which are grooved to fit the surface of a section of extra heavy wrought-iron pipe, upon which they roll. From this tightener the rope passes direct to the right-hand groove of the pulley on the main shaft above the engine, the tightener-pulley being inclined sufficiently to make the bottom come in line with the left-hand, while the top comes in contact with the right-hand groove.

The main pulleys which drive the shaft above the engines are mounted upon and keyed to sleeves 10 inches diameter, which extend out on each side far enough to form journals, by which they are supported in pedestals independently of the shaft (see Fig. 33). Through the sleeve is a hole considerably larger than the shaft which passes through them and which is supported by separate

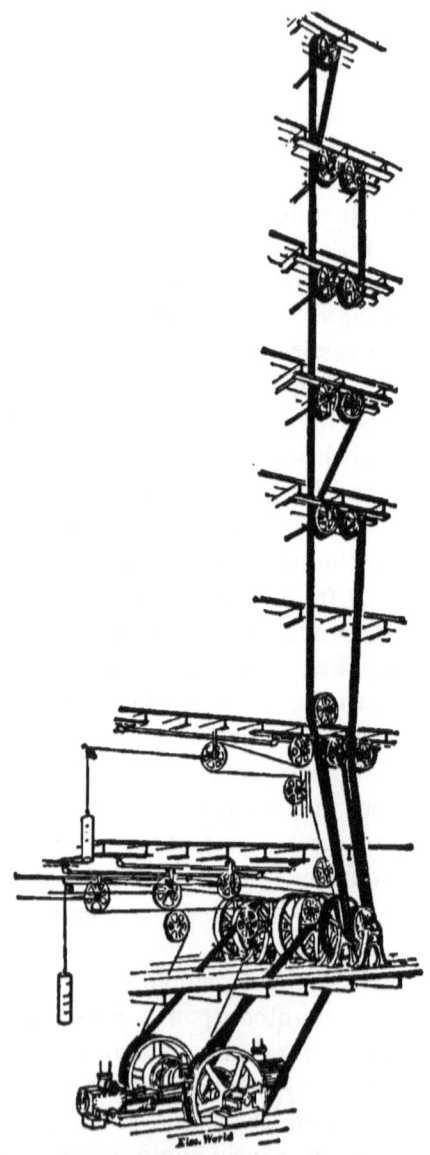

FIG. 10.—AMERICAN OR WOUND SYSTEM OF ROPE TRANSMISSION.

pedestals. One end of each sleeve is so formed as to make, in connection with a sliding collar which is on the shaft, a positive interlocking clutch, which can be thrown in or out by a lever. In this way both engines may be working at the same time, or the shaft may be run by either engine alone, the other pulley standing and imposing no friction upon the shaft. All the bearings of this shaft are adjustable laterally by set-screw and vertically by wedges. The other large pulleys upon this shaft are driven by friction-clutches, and are used for driving dynamos, each pulley driving two dynamos arranged tandem, one belt running over the other. This shaft runs at 220 revolutions and is $4\frac{1}{2}$ inches diameter. At one end of the main shaft is a pulley having twelve grooves, in which run two ropes side by side to the top of the building and around the various pulleys down again. Either of these ropes is calculated to be amply strong for the work, but two are used to avoid the necessity for stopping should one break. Each of them winds three times around the pulleys, thus giving six driving-strands. To avoid crowding, the tightener for one of these is placed upon the floor above the other. The ropes pass from the main pulley three times around the pulley above and then go to the upper floors, as indicated. From the shafts, the wood-working machinery, blowers for the foundry, and some of the elevators are operated; the other lines being used for driving light machine tools, such as are used in making electrical apparatus.

Each floor has a cut-off coupling, which is so arranged that in case of accident it can be cut off at a minute's notice, or when running overtime any floor can be cut off, thus saving the cost of running any more machinery than is necessary.

The use of the tension-carriage plays an important part in the American system of rope transmission. As usually made it is automatic in its operation and so weighted as to give a constant tension to the rope, as indicated in Fig. 11. In this arrangement an initial tension is given

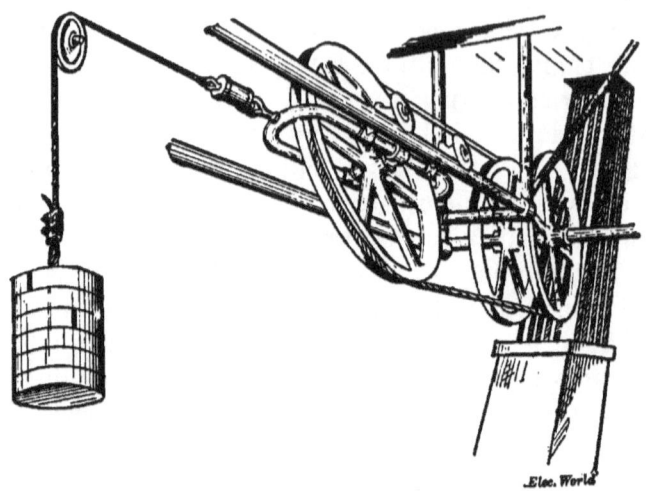

FIG. 11.—AUTOMATIC TIGHTENER FOR ROPE TRANSMISSION.

and maintained by the automatic tension-carriage, which is free to move backward or forward on a horizontal track

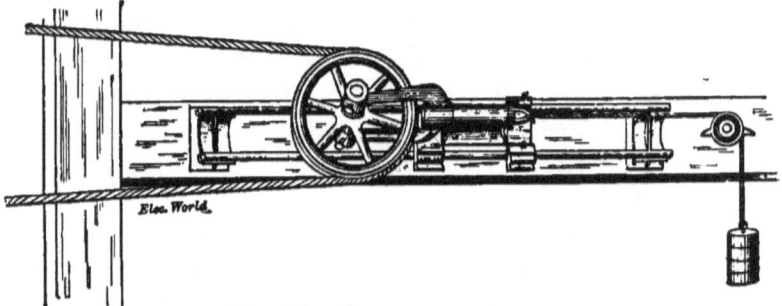

FIG. 12.—TENSION-CARRIAGE.

as the load changes or the rope stretches, always taking up the slack and maintaining the proper tension. Another form of tension-carriage is that shown in Fig. 12.

In this case the carriage is mounted on gas-pipe or solid shafting, and is provided with ball bearings arranged with cast pockets so that the balls are allowed to circulate lengthwise of the bearings.

An arrangement of light angle-iron tracks supporting a four-wheeled carriage is used to a considerable extent and makes an excellent tightener where it can be employed, as it is cheap and readily set up.

It is obvious that vertical tensions may be arranged in a similar manner to those shown in Figs. 11 and 12. In such cases the weight may be suspended directly from the carriage or even the sheave, as in Fig. 13, or it may be led off in any desired direction either above or below the tightener-pulley. The tightener pulley is often inclined from the vertical, so that its projection is equal to the width of the driving pulley, in which case it not only serves to maintain a constant tension in the rope, but it thus acts as a guide to conduct the rope from one groove to another.

FIG. 13.

FIG. 14.—ROPE-TIGHTENER.

A modification of the usual belt-tightener is sometimes used for rope drives, as shown in Fig. 14, but this, it will be noticed, is positive in action and is only used to increase the tension on the ropes as the latter become extended with use.

In the same way the tightener shown in Fig. 15 * is used to take up any slack that may occur in the rope.

In this case the tightener-pulley T is mounted on a standard free to slide in the bottom guides G. A weighted lever L is connected to the pinion-shaft S by means of a ratchet-wheel and pawl (not shown in the figure). On the other end of this shaft and rigidly connected to it a pinion meshes with the rack R upon one of

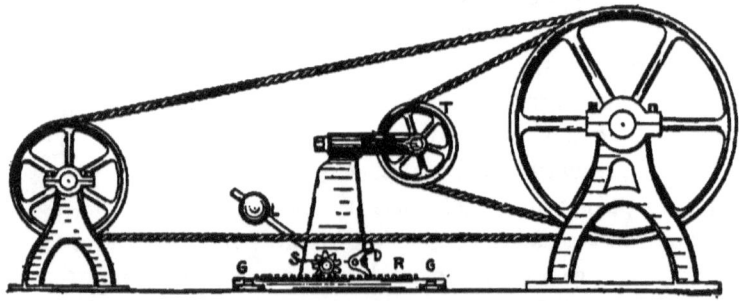

FIG. 15.—DYBLIE'S ROPE-TIGHTENER FOR DYNAMOS.

the guides and causes the standard and tightener-pulley to move along the base and thus automatically take up the slack as it occurs. A detent, D, in the carriage catches in the teeth of the rack and prevents the tightener from slipping back. It is evident that any form of adjuster which will not allow the tension-pulley to move in both directions—either back and forth or up and down—does not maintain a uniformity of tension; as the humidity in the air may cause a rope to shrink very materially in a short time the tension on the rope will be greatly increased if the tightener-pulley is prevented from moving in.†

---

* Patented by J. A. Dyblie, Jan. 14, 1890.

† Mr. Louis I. Seymour states that in a certain out-door drive at

It is a well-recognized fact that atmospheric changes affect the length of a rope, which in the presence of moisture always contracts. Experiments have shown that a dry hemp rope 25 feet long will shrink to 24 feet upon be-

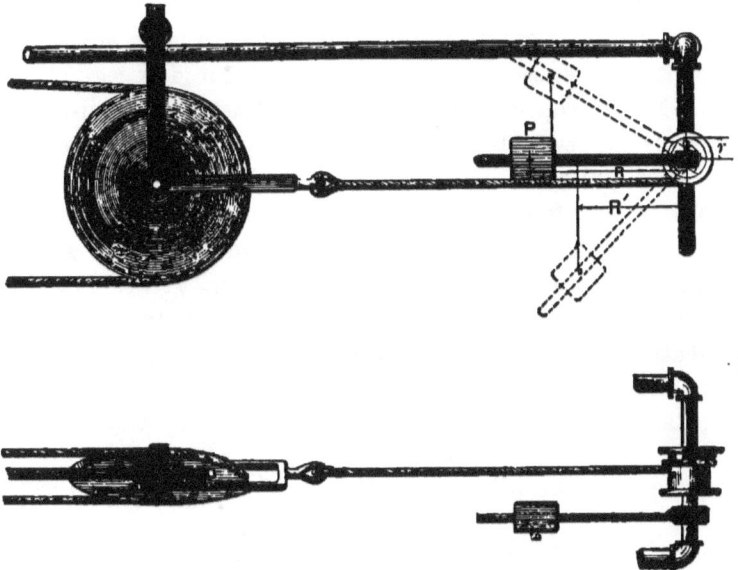

FIG. 16.—TENSION-CARRIAGE WITH VARIABLE ARM.

ing wet.* It is for this reason that, in addition to taking up the slack caused by variation in load, provision should

---

the Plymouth Cordage Company's Works in which 1¼-inch rope is used to transmit 125 h. p. the tension carriage is drawn in about eight feet by the shrinkage of the ropes during a severe storm; that is, the rope is shortened about sixteen feet. The total length of rope in this case is approximately 1600 feet, so that the contraction is thus about one per cent of its total length. The rope used was of superior quality manilla, laid with plumbago and tallow, otherwise the shrinkage would have been greatly in excess of the amount stated.

* *Indian Engineer*, 1888.

be made for maintaining a proper tension in the rope when the latter is variable in its length, due either to atmospheric changes, or permanent elongation as the rope loses its elasticity. This variation in length is particularly noticeable in rope-drives which are subjected to exposure from the weather.

An arrangement sometimes adopted with horizontal carriages as a substitute for pulley and hanging weight is shown in Fig. 16. In this arrangement the tail-rope, usually of wire, is wrapped around and secured to a grooved pulley-sheave keyed to a shaft which is fixed in position but free to rotate. A weighted lever is secured to the shaft by means of a set-screw and maintains a tension on the rope by virtue of its moment. It will readily be seen that this tension will not be constant, for the effective lever-arm of the weight, and hence the pull on the tail-rope, will vary with the position of the lever; thus in its normal position with the lever horizontal the tension in the tail-rope will be $T = \dfrac{PR}{r}$; but if the tension-carriage moves either in or out on its guides the lever will assume a new position as shown in dotted lines, in which case the tension will now be $T = \dfrac{PR'}{r}$.

As the initial tension which gives adhesion to the slack side of the rope varies with the weight supported by the tension-carriage, it is obvious that an increase of this weight will increase the power which may be delivered by the rope. As, however, the horse-power is proportional to the difference in stress in the driving and slack sides of the rope, the less weight on the tightener consistent with obtaining sufficient frictional resistance to slipping, the better will the ropes work.

An example of rope-driving, in which tension-carriages

are arranged to work vertically, is shown in Fig. 17. In this case the engine develops about 45 horse-power and runs at 90 revolutions per minute, corresponding to which the velocity of rope is about 1700 feet per minute. The fly-wheel is six feet in diameter, and is grooved for five $1\frac{1}{4}$-inch ropes. The main sheave on the jack-shaft is also 6

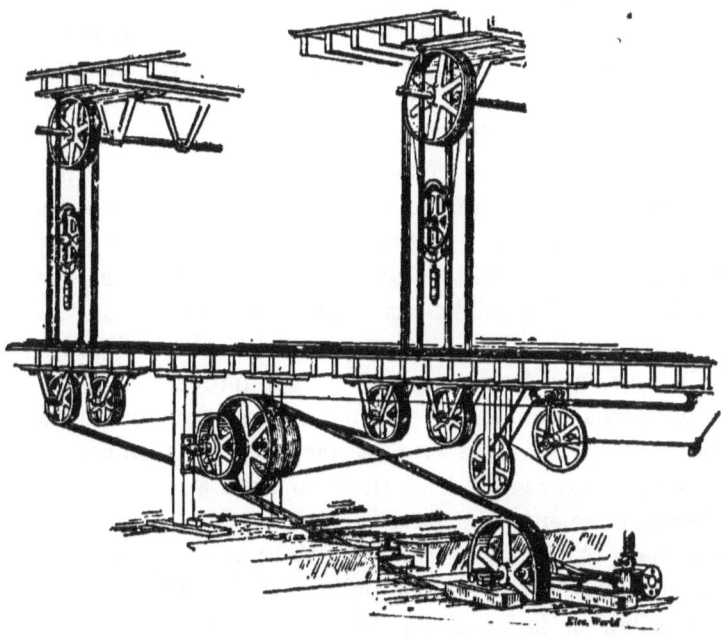

FIG. 17.—ROPE-DRIVE WITH VERTICAL TENSION-CARRIAGES.

feet in diameter, and is grooved for five ropes. The rope passes continuously from the fly-wheel to the main sheave, making five wraps; then over the deflecting-sheave to the horizontal tension-carriage, and back to the fly-wheel.

The jack-shaft sheaves are grooved for four 1-inch ropes, and are each provided with a friction-clutch, giving the line two shafts and, practically, the advantages of independent motors, so that in case of accident to either the other can be run independently.

The sheaves on these shafts are five feet in diameter, and are grooved for five 1-inch ropes,—four wraps being used for the transmission of power; the other groove is for the return from the vertical tension-carriage, which is shown just beneath the pulley on the line-shaft.

Although used extensively for main drives and in many cases for intermediate driving, ropes have not come into use for general work throughout the factory. In certain cases the whole of the driving is done by ropes, no cross-shafts, gears, nor belts being used; but the greater convenience of flat belts for conveying power to the machine in ordinary shop-transmissions is so thoroughly acknowledged, that any attempt to substitute rope for belting in general would be met with little favor; nor would this be practicable under the conditions which now obtain in our mills and factories. One reason for this is the difficulty in shifting a rope from a tight to a loose pulley.

Numerous devices have been employed for this purpose, and in some few cases with satisfactory results; but the system does not readily lend itself to such work, nor would the arrangements that have been adopted be generally applicable. Another reason for its non-employment for general work is the size of pulleys, and distance between shafts which is necessary for satisfactory working with ropes, and which, obviously, would exclude its use in many instances.

Where the power is transmitted to a shaft whose axis is in a different plane from that of the driver, it is often inconvenient or impracticable to use bevel-gears or universal couplings. For such special cases rope-driving is particularly satisfactory, as a flexible rope will readily conform to any direction, and transmit its full power when arranged with suitable idlers.

ROPE-DRIVING. 35

An example of such transmission is given in Fig. 18. This drive transmits 150 h. p. for the main shaft at *a* to another at *b*, whose axis makes an acute angle with the first, and which is several feet lower. The drive-sheave *a* is 5 feet 3 inches in diameter and runs at 160 revolutions per minute; it is grooved for six $1\frac{1}{4}$-inch ropes, which have a velocity of 2800 feet per minute; *c* and *d* are idlers, the faces of which are nearly parallel to the drive-sheave; *e*

FIG. 18.—ROPE-DRIVE WITH SHAFTS AT AN ANGLE.

and *f* are double idlers, there being two sheaves on each shaft, one $1\frac{1}{2}$ inches less in diameter than the other. This arrangement of idlers in quarter-twist drives of this class has been introduced where there are more than four ropes in the system, with the object of reducing the wear consequent to the friction produced by the side lead of the rope. In drives where there are more than eight ropes a cone has been used to a limited extent for the same purpose.

Such a practice is, however, very unsatisfactory, as the trouble encountered by the differential driving exerted by each rope on a different diameter of the cone is greater than that which it attempts to obviate. However, by making the several steps on the cone separate, so that they form a series of independent idlers of different diameters,* the difficulty is overcome.

It is evident that the employment of multiple idle sheaves of equal diameter possesses many advantages over a multiple-grooved pulley when used as guide-pulleys.

In this case each sheave is independent of the others, and thus prevents in a large measure the evils due to differential driving and slip which would otherwise occur with fluctuations of load.

With ordinary transmissions where the vertical distance between shafts is as great as 100 times the diameter of rope no trouble is experienced from the side lead of the rope, and, usually, no provision is made to obviate it.

An application of rope-driving to shafts at right angles, embodying several excellent features, is shown in Fig. 19.

The plant is designed to transmit 250 horse-power from a 14-ft. fly-wheel, a, which is grooved for twelve 1-inch ropes. The line-shafts, m and n, are driven independently, and each drive has its own tension-carriage. The rope-sheave b is 72 inches in diameter, and is grooved for five ropes. At the side of the 72-inch sheave is a single-grooved idler, i, loose on the shaft, which serves as a guide for the rope to the tension-carriage.

The substitution of a loose idler for an extra groove on the driven pulley in rope transmissions is due to Mr. Spencer Miller, although it has long been in use in cable-railway

---

* This feature is the subject of a patent granted to Mr. John Gregg, March 11, 1890.

practice; the advantage in its use is evident when we consider that the tension-carriage, drawing out the stretch of the rope, must necessarily drag the first rope through the

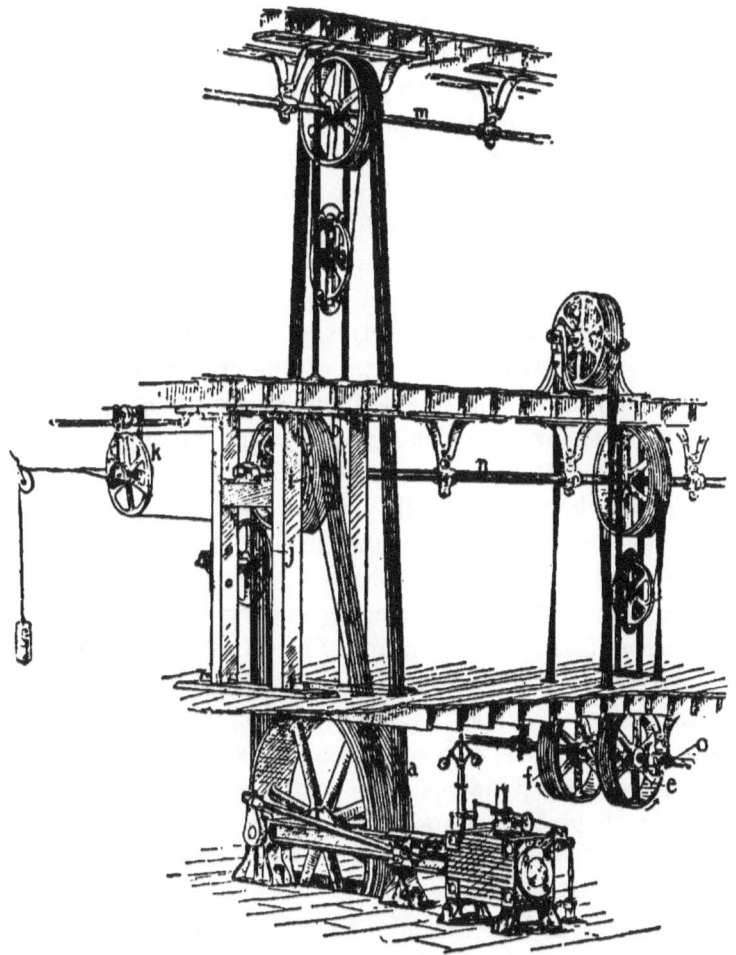

FIG. 19.—TRANSMISSION AT RIGHT ANGLES.

groove of the pulley, which will require an excessive weight on the tightener pulley and a greater length of time before equilibrium is restored. By having this groove made into an individual wheel free to rotate on the shaft, this difficulty is overcome, and the transmission responds very freely to changes of load, so that when heavy machines are thrown on and off the ropes are not set in vibration, but the tension-carriage sheave K slides back and forth on the track, taking up the shock, with a minimum amount of wear on the ropes.

In the present case the rope runs continuously around the fly-wheel and sheave from groove to groove. As it leaves the fly-wheel at the left hand it passes over the idler i to the tension-carriage sheave K, which is suspended on adjustable hangers from a single-pipe track. This sheave is tilted by means of the adjustable hangers, so that the top is in line with the centre of the groove of the idler, and the bottom is in line with the centre of the groove of the guide-sheave j, which serves to carry the rope back to the right-hand groove of the fly-wheel.

The employment of multiple idlers instead of a multiple-grooved idle pulley is also shown in this figure, where power is transmitted from the shaft n to another o at right angles to the first. By engaging one or the other of the driven pulleys e or f by means of a clutch, the shaft o may be driven in either direction.

More recently ropes have been introduced to drive dynamos and special isolated machines, and where the distance between dynamos and engine or driving-shaft is sufficiently great to allow a moderate sag in the ropes, such drives have been found to work very satisfactorily, provided other conditions are favorable. Where the distance between shafts is limited, more wraps should be given to the rope in order to lessen the tension in each member. One great fault with dynamo drives is the use of too small a pulley on the armature-shaft. We can point to a score of plants using rope transmissions from a jack-shaft to dyn-

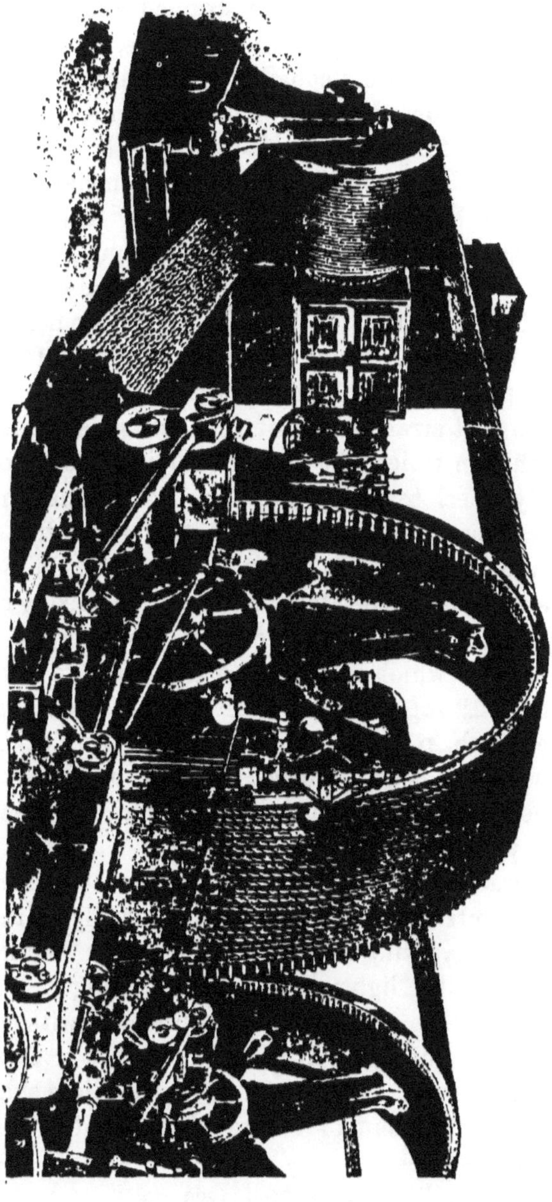

ROPE-DRIVING. 39

FIG. 20.—ENGLISH SYSTEM OF ROPE-DRIVING APPLIED TO DYNAMOS.

amo in which the rope is overstrained and the dynamo pulley is only half as large as it should be, in consequence of which the ropes are a constant source of trouble.

With cotton ropes the pulley may be somewhat smaller than that used for a similar size of manilla rope; but in any case there is a certain minimum diameter of pulley which should be used for any given rope. (See page 179.)

Where the required number of revolutions cannot be attained with the size of pulley imposed by the various conditions, if the jack-shaft cannot be speeded up nor a larger driving-pulley used, in such cases it would be better to take out the rope and put in a good leather belt.

The simplest arrangement of rope transmission for dynamos is that in which the rope is carried direct from the engine fly-wheel to the grooved pulley on the armature shaft, as shown in Fig. 20, which represents the system used in the station of the Liverpool Overhead Railway.*

There are four horizontal compound condensing engines with cylinders $15\frac{1}{2}$ and 31 inches in diameter, 36 inches stroke, each of which is connected to a separate generator by means of 19 ropes $1\frac{1}{4}$ inches diameter. Each engine is rated at 400 h. p. when running at 100 revolutions per minute with 120 pounds initial pressure; as the fly-wheels are 14 feet in diameter, the rope velocity will thus be about 4400 feet per minute.

In general it is more desirable to drive the machines through an intermediate jack-shaft, especially so in those cases where a varying amount of current is required, as, for instance, in the lighting of public buildings. Such an arrangement, when the jack-shaft is provided with suitable friction or jaw clutches, will permit machines to be thrown on or off as desired. The use of an intermediate shaft also permits the attainment of the requisite speed of the dyn-

---

* *Power*, May, 1893.

amo with moderate proportions of pulleys. In fact with many of the smaller engines in use a jack-shaft is essential if we wish to use rope-driving.

Take, for example, a 75-h.-p. Corliss engine, running at 85 revolutions per minute; the diameter of fly-wheel for this engine is 10 feet, and if we wish to drive the dynamo direct through a ¾-inch rope, the pulley on the armature-shaft should be, preferably, not less than 24 inches diameter; the speed of the armature would then be only 425 revolutions per minute. To obtain a suitable speed, the driven pulley could not be more than about 12 inches in diameter, and with larger ropes this difference would be still more pronounced.

A similarly rated high-speed engine runs at 230 revolutions per minute, and is provided with a fly-wheel 5 feet in diameter; with the same 24-inch pulley on the dynamo the speed of the latter when driven direct would be not more than 570 revolutions per minute, so that in this case also the driven pulley would have to be reduced very much below that size which has been found best adapted to the work.

It is true that the diameter of driving-wheel on the engine-shaft could be increased, and this is sometimes done. In the cases quoted the diameters necessary to give the required speed would be about 20 feet for the Corliss and 7 feet for the automatic engine. As these sizes give a circumferential velocity within the limit of safety from the action of centrifugal force of the metal in the rim, it would be highly desirable to use such driving-wheels if other practical considerations did not preclude their use. A driving-wheel 7 feet in diameter could readily be used on the high-speed engine without materially augmenting the journal friction, and the increased rim speed would be a beneficial factor in preventing momentary fluctuations due to change in load.

With the Corliss engine, however, the increased weight due to a large built-up fly-wheel 20 feet in diameter would usually debar its use on an engine of this capacity; in addition to this the large diameter would prevent its use in many locations, even if the increased weight and loss of power were no hindrance. Under these conditions the use of a jack-shaft is the most suitable arrangement.

In many cases it is desirable to use two engines so arranged that either or both may furnish the power to any one of several dynamos, as shown in Fig. 21, which represents a 175-h.-p. rope-transmission plant erected by the Link-belt Engineering Company in the Virginia Hotel, Chicago, where two Corliss engines are each connected to a jack-shaft, having five counter-drives to the dynamos. Both the driven and driving sheaves on the jack-shaft are loose on the shaft, and are connected to it by means of friction or jaw clutches, thus permitting either or both engines to be run, or any one of the five dynamos to be thrown in or out of use. The positive jaw-clutch is used on the driven sheave, as it does not readily get out of order and is preferred by many engineers to the average friction-clutch, especially in those cases where much power is transmitted.

If it is desired to couple one engine to the shaft while the other is running, the former is speeded up until the loose driven sheave comes up to the speed of the shaft, when the dog-clutch may be readily thrown in gear without shock. With this arrangement there is a strong tendency for the bearings of the driven sheaves to heat when not coupled to the shaft; for this reason provision should be made to reduce the friction between the loose pulley and the shaft by relieving the tension on the rope when not in use, or, what is much better, the loose sheave should be mounted upon a hollow sleeve supported in pedestals independently of the shaft, as noted in description of plant shown in Fig. 10; see also Fig. 33. In this way either pulley may be at rest and impose no friction on the shaft.

FIG. 21.—ROPE TRANSMISSION WITH JACK-SHAFT.

If T_1 is the maximum stress in a rope, T_2 the stress in the slack part, and $P (= T_1 - T_2)$ is the driving force, then, as we shall show subsequently, the ratio of the maximum stress T_1 to the driving force P, or $\frac{T_1}{P}$, will vary from about 1.5 to 5.5—depending upon the speed of rope, the coefficient of friction, and the angle embraced by the rope on the circumference of the pulley. In order, then, to have the transmitting force P as large as possible for a maximum tension T_1, the tension in the slack part of the rope necessary for adhesion must be reduced to a minimum. To a certain extent this can be obtained by decreasing the angle between the sides of the groove, but if carried too far this is a detriment rather than an advantage, for if the angle is sufficiently acute the rope will wedge and require more or less force to pull it out of the groove. The remaining expedient is to increase the arc of contact.

It can be shown that the friction of a cord or rope wrapped upon a fixed cylinder is independent of the diameter of the cylinder, and that it increases very rapidly with an increased arc of contact.* If the conditions are such that the coefficient of friction $\phi =$ one third, a tension of one pound at the end T_2 of the rope, Fig. 22, will support a strain at the end T_1 of:

1.69	pounds for an arc of contact equal to				¼	coil.
2.85	"	"	"	"	"	½ "
8.12	"	"	"	"	"	1 "
65.94	"	"	"	"	"	2 "
530.43	"	"	"	"	"	3 "
4,348.56	"	"	"	"	"	4 "

Therefore by increasing the number of wraps around the cylinder it is possible to increase the difference between the tensions in each part of the rope almost indefinitely. It will be noticed here that the rope is not in flying motion,

---

* Weisbach, vol. I. p. 360.

which would cause an equal centrifugal force to be set up in each member, thus altering the ratio of stress. As the centrifugal force varies with the square of the velocity, there is with an increasing speed of the rope a decreasing useful force and an increasing total tension on the slack side; but up to a given limit, which we shall subsequently show lies between 4000 and 6000 feet per minute, the total power transmitted for a given maximum tension in the rope will increase with the velocity.

The great advantage thus obtained by increasing the adhesion was very early applied to numerous mechanical devices, chiefly for winding and hoisting purposes, and later for haulage systems. We are indebted to Willis* for the following quaint account and sketch (Fig. 23) of an arrangement for obtaining a continuous motion with a continuous travelling-coil, first suggested by the author of the article, Sir Christopher Wren, over two hundred years ago.

"A DESCRIPTION AND SCHEME OF DR. WREN'S INSTRUMENT FOR DRAWING UP GREAT WEIGHTS FROM DEEP PLACES."

Read May 5, 1670.

"Having considered, that the ways hitherto used in all Engins for winding up Weights by Roaps have been but two, viz. the fixing one end of a roap upon a cylinder or Barril, and so winding up the whole coyle of roap ; the other by having a chain or a loose roap catching on teeth, as is usual in clocks: but finding withall that both these wayes were inconvenient the first, because of the riding of much roap in winding one turn upon another; the other, because of the wearing out of the chain or roap upon the teeth, I have, to prevent both these inconveniences, devised another to make the weight and its counterpoyse bind on

---

* "Principles of Mechanism," p. 429 et seq.

the cylinder, which it will doe if it be wound three times about.

"But because it will then in turning, scrue on like a worm, and will need a Cylinder of a very great length, therefore if there be two cylinders each turned with three notches and the notches be placed alternately, the convex

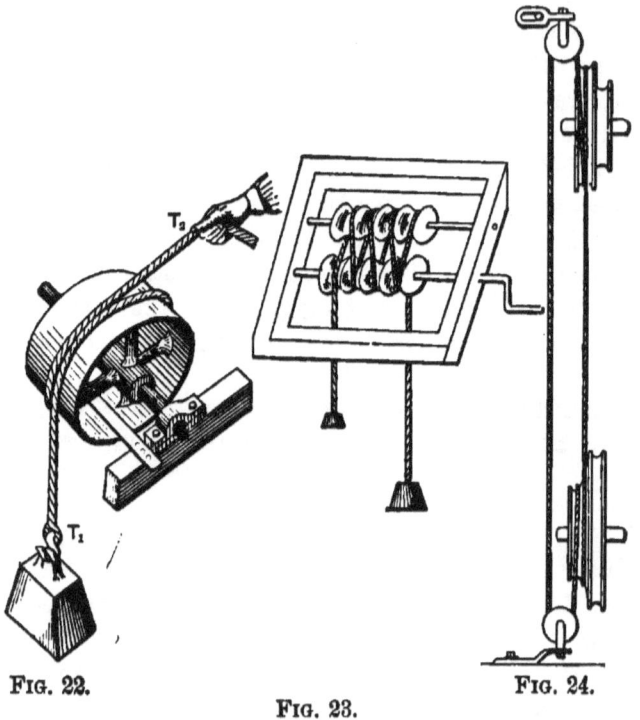

FIG. 22.   FIG. 23.   FIG. 24.

edges to the concave as in the figure here adjoyned, the roap being wound three times about both cylinders, will bind firmly without slyding and work up the weight with a proportionable counterpoyse at the other end of the Roap."

The method of obtaining increased adhesion for a given wrap by increasing the number of coils in contact with each pulley has long been in use in rope-driving. In some of the earlier applications the grooves of the pulleys were semicircular in section, and of sufficient size to allow the rope to embrace the entire circumference of each pulley as represented in Fig. 24.*

A better arrangement is that shown by Overman,† in which the advantage of the angular groove is obtained, and the ropes are not worn by rubbing against each other. This is shown in Fig. 25, and is thus described:

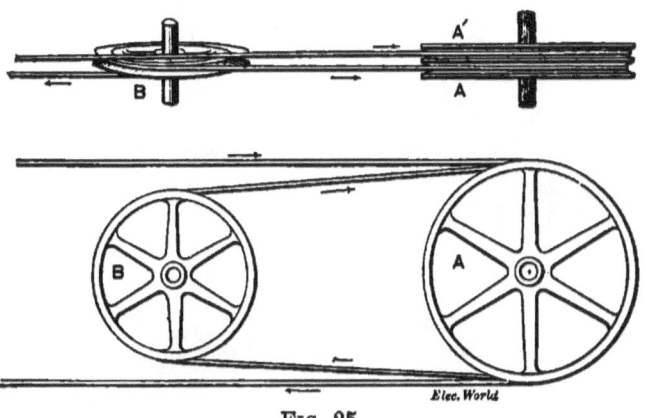

FIG. 25.

"If the pulley A is grooved, of which at least two are fastened to the same shaft, the rope is directed on one of these pulleys, and passing around it goes to B, which revolves on an inclined axis, such that the rope will be received from A' and delivered to A in the plane of the grooves. The number of pulleys may be multiplied to gain adhesion. This method of augmenting friction is preferable to the tension-roller, as no increase of tension is required; and it has the additional advantage of bending the rope in the same direction, which makes it more durable."

---

* Willis. Principles of Mechanism.
† Overman's "Mechanics," 1851.

A similar arrangement was introduced in the San Francisco Cable Railway in 1877. In this case, Fig. 26, the endless hauling-cable is passed alternately backward and forward over two grooved drums a sufficient number of times to obtain the necessary driving adhesion.

Positive motion is imparted to the drum *A*, which hauls in the cable, while a second pulley, *B*, acting as an idler,

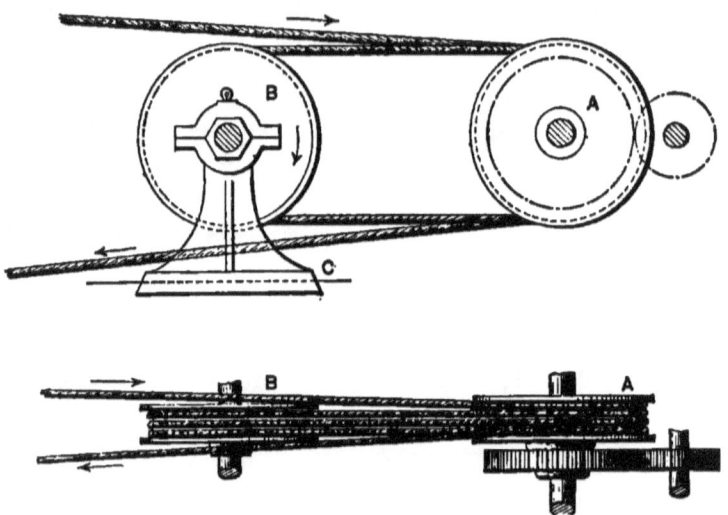

FIG. 26.—COIL-FRICTION, STREET-RAILWAY CABLE.

increases the frictional grip of the cable on the drum, by virtue of the increased arc of contact due to the number of wraps.*

When, however, the cable permanently lengthens by stretching, the drum *B* may be moved further back by means of the sliding-base *C* so as to take up the resulting slack.

By the use of this winder pulley the property of frictional adhesion produced by successive coiling is perfectly

---

* "Cable or Rope Traction." J. Bucknall Smith, C. E.; *Engineering*, London, 1887.

effectual, for, although each coil is only in contact with a semi-circumference, the accumulation of frictional resistance is produced precisely as if entire circumferential grooves were employed.

However desirable such winder pulleys may be for cable haulage or hoisting purposes, their advantage is greatly overestimated when applied to continuous rope-transmission. A little consideration will show that the frictional adhesion produced by a tension weight acting on a running rope with numerous wraps is entirely different from coil-friction. In the latter we have an accumulation of friction by which a small resistance applied at one end of the rope is able to hold an enormously greater load at the other end.

In the continuous-rope system of power transmission, however, the load is distributed among all the wraps, so that when properly adjusted each wrap carries an equal proportion of the load and is subjected to an equal resistance on its slack side. There are few cases where the combination of the winder-pulley with the continuous-rope system offers any decided advantage over other methods.

There is an incidental advantage in using a winder, especially in those cases where the difference in diameters between the driver and follower is quite appreciable; under such conditions the adhesion of the rope on each pulley may be made more nearly uniform by the employment of a winder, and there is less liability of the rope slipping in its groove, but this may usually be obtained more satisfactorily by other means (see page 168).

In numerous cases ropes using winder-pulleys have been installed without regard to the work to be done or strain put upon the ropes, and many of the evils of rope-driving are directly traceable to this cause.

Many engineers are opposed to using the winder-pulley

in any form whatever, but occasionally it may be used to advantage.

In outdoor or long-distance transmissions, or in special cases where it is desired to transmit a maximum power without undue stress in the rope, or in particular cases where the ordinary working stress may be exceeded, a winder-pulley may frequently be used to advantage, if the slack-side tension be reduced accordingly. Using a winder-pulley and increasing the back tension will permit a very large increase in the power transmitted; but since this imposes an excessive strain on the rope, it soon wears out, and is a constant source of trouble.

The gain in power by increasing the adhesion will be at the expense of journal-friction, which is thus augmented by the employment of wider-faced driving and driven pulleys, in addition to that due to one or two more winder-pulleys; the wear of the rope, both external and internal, will also be greatly increased on account of the greater number of flexures given to the rope in passing over the winder-pulleys.

The use of a winder-pulley at each end of a long drive in which only a single strand runs from the driver to driven pulley is an example of the application of coil-friction to the continuous-rope system; in this case both the working load and the back tension is carried by one rope instead of being distributed among several wraps, as usually happens in this system.

That the percentage of gain is not as great as might be expected from the employment of coil-friction, will be seen from the following considerations:

The ratio of tensions in the tight and slack sides of a rope running over two pulleys is dependent upon several factors, and may be determined from

$$T_1 = T_2 (\epsilon^{\phi a}),$$

in which T_1 = tension in tight side of rope,
T_2 = tension in slack side of rope,
ϵ = base of hyp. log = 2.7183,
ϕ = coefficient of friction,
α = arc of contact (circular measure).

In this case the influence of centrifugal force is neglected.

Since the power transmitted by a wrapping connector is dependent upon the difference of tensions in the tight and slack sides, it is evident that with an assumed total tension T_1 the available force for transmitting power will increase as T_2 decreases.

But T_2 may diminish as $\epsilon^{\phi\alpha}$ increases, that is, since ϵ is a constant and ϕ is constant for a given pulley and material, T_2 decreases as α increases; hence if we increase the arc of contact the tension in the slack side of the pulley may be decreased in a ratio greater than unity, depending upon the factors involved; in which case the net force P available for transmission will be increased, while the original assumed allowable tension remains the same.

For example, if T_1 = 200 pounds, $\phi = 0.3$, $\alpha = 2.88$ ($\alpha° = 165°$), we shall have

$$T_2 = \frac{200}{(2.718)^{0.3 \times 2.88}} = 84,$$

and the net force $P = 200 - 84 = 116$ pounds.

Under the same conditions, if we pass the rope from the driver over a winder-pulley back and forth twice, and then to the driven pulley and its winder in the same way, we should be able to transmit a little more than one and a half times as much power at the same speed without increasing the working tension in the rope.

In this case $\alpha = 12.56$ and $\dfrac{T_1}{T_2} = 43.1$; hence

$T_2 = 4.6$ and $P = 200 - 4.6 = 195 +$ pounds.

$$\therefore \frac{\text{Horse-power in second case}}{\text{Horse-power in first case}} = \frac{195}{116} = 1.7.$$

Now, if it were practicable to maintain the same slack-side tension T_2 in these two instances and increase the driving-side tension under the conditions of the second case, viz. $\frac{T_1}{T_2} = 43.1$, we should have

$$P = T_1 - T_2 = (43.1 \times 84) - 84 = 3536 \text{ pounds},$$

and the ratio of power transmitted will now be $\frac{3536}{116}$, or thirty times as great as before.

These results indicate that while we may vastly increase the driving-side tension for a given slack-side tension, by using a winder in the manner indicated, that is, with a single wrap connecting driver and follower, yet if we wish to maintain an assumed maximum working tension for a given-sized rope the percentage of gain will not be very great under the usual requirements of rope-driving.

For a temporary drive the working strain may be increased to about twice the usual value, but for a permanent installation the usual working value should not be exceeded.

Where a number of ropes are employed on a short-drive it is questionable whether the winder-pulley possesses sufficient advantages to warrant its employment in place of the continuous-wrap or individual-rope systems. In any case the conditions should be carefully considered, and the actual gain compared with the various losses involved.

A recent example illustrating the application of the winder-pulley is shown in Fig. 27, which represents the system of rope-driving installed by Messrs. Hoadley Bros. in the Fifty-second Street electric power-house of the Chicago City Railway Company.

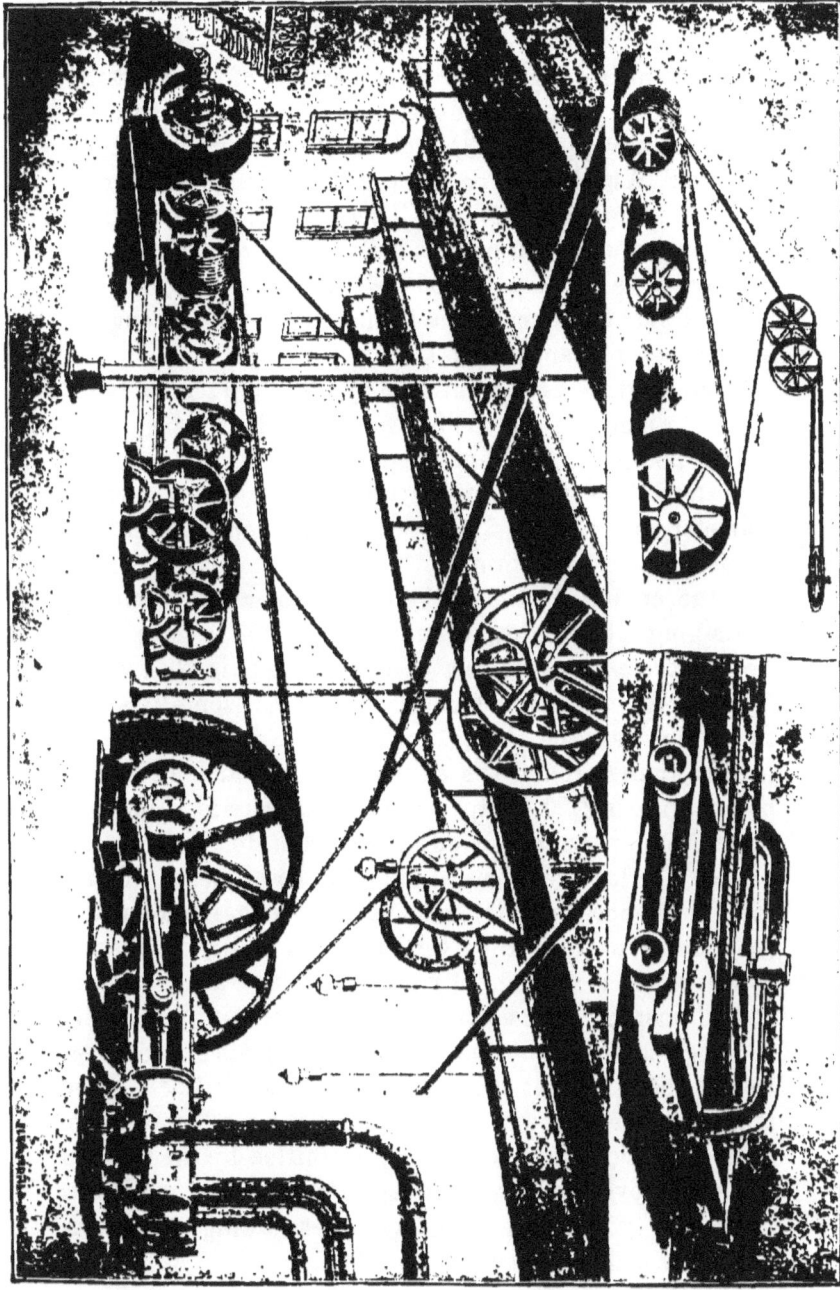

The plant is designed for 10 generators of the Westinghouse No. 6 type, running 300 revolutions per minute. There are also to be ten 24-inch by 48-inch engines of the improved Wheelock type, arranged in five pairs, two of which are now in operation. These run at 100 revolutions per minute with 100 pounds boiler-pressure. The power usually varies from about 200 to 1000 h. p., but during the heavy traffic throughout the summer each pair of engines has frequently transmitted 1500 h. p.

These engines have a built up fly-wheel (10 segments) 18 feet in diameter, 39 inches face, which weighs about 50,000 pounds. The rim is grooved for 21 wraps of $1\frac{1}{4}$-inch manilla rope. The driven pulleys are 6 feet in diameter, and contain 32 grooves for the rope, which runs about 5600 feet per minute. Between the driven pulleys and the engine fly-wheel there is placed a 6-foot winder, containing 11 grooves, around which the rope is carried before passing to the tension-sheave (Fig. 28), which in the present arrangement is placed horizontally above the engine near the ceiling, as shown in Fig. 29. Thus the rope is wound around the engine fly-wheel and the driven pulley, making 20 wraps; then it is carried from the driven pulley to the winder back and forth 11 times, thence it is led over vertical guide-pulleys, 7 feet in diameter, to the horizontal tension-sheave 54 inches in diameter, then down over another vertical guide-pulley to the fly-wheel, where it started. By this means the arc of contact of each member of the driving-rope is increased practically 180 degrees when all the ropes have adjusted themselves to the load, so that the power transmitted with the same tension in the rope will be about forty per cent more, if we neglect friction, than would be transmitted by the twenty-one wraps over the fly-wheel without the use of a winder.

The net gain will be considerably less, owing to the various losses which this system entails.

ROPE-DRIVING. 55

Fig. 30.—Rope Transmission from Horizontal Turbine.

The application of ropes to transmit the power from a water-wheel to a line-shaft 40 or 50 feet or more above the axis of the wheel has lately received considerable attention, and offers many advantages over the ordinary method in which a vertical shaft is used. The extreme weight of the latter in many cases makes it a difficult matter to provide a suitable bearing to support it. In such cases a horizontal turbine is used, and the wheel-shaft carrying the rope-pulley is extended and suitably supported, as shown in Fig. 30.

The station of the Brush Electric Light and Power Co. at Niagara Falls is driven in this way.

A line-shaft runs through the building, with one end extending over the wheel-pit; to this are belted the generators in the usual manner. Seventy-five feet below this shaft is located a 15-inch horizontal Victor turbine in a case of boiler-iron, its shaft extending to bearings supported by bridge-trees, which in turn are carried by the foundation I beams that support the wheel-case. This shaft carries an iron pulley 40 inches in diameter, grooved for 12 ⅞-inch manila ropes. (Cotton was tried, but was not satisfactory here.)

The pulley on the driven shaft above is of wood, 70 inches in diameter. The driving side of the ropes hang perpendicularly, and are free from the driver to the driven pulley.

The slack side has two idlers or guide-pulleys, one of which is situated immediately below the driven pulley, and the other is about 20 feet above the driver. A tightener is adjusted in a running frame, in line with the driven and upper guide-pulleys.

In putting on the rope the following course is taken: "Commencing at one side of the pulleys the rope is passed around from the driver to the driven pulley in every alternate groove until the opposite side is reached, thence around the tightener-pulley in the running frame, which is hung on an incline in such a manner that its discharging side is in line with the side of the driver pulley whence we started. The remaining grooves are then filled, and the ends of the rope are spliced in position around the idler. Thus it is readily seen that there are two strands of the rope on the idler at all times. The object of this is to have one solid piece of rope on the idler at the same time that the splice is, so as to relieve the spliced piece of the strain of the idler. This system is giving far better satisfaction than has the upright shaft to those who have tried both."*

The plant designed by Mr. Robert Cartwright for the electric station of the Citizens' Light and Power Company of Rochester, N. Y., is worthy of careful study, and may be considered a representative modern plant, adapted to use steam or water-power, and employing both ropes and belting. †

Fig. 31 represents a cross-section of the station, and shows the general arrangement of the plant. The waterwheels are twin Poole-Leffel central-discharge turbines, 23 inches in diameter, and at a speed of 560 revolutions per minute, under a head of 92 feet 6 inches, develop 500 horse-power each, with a discharge of 3800 cubic feet of water per minute. The wheels proper are made of phosphor-bronze, with buckets of Otis steel, tinned. The wheel bed-plates are heavy cast-iron box sections, machined and bolted together with heavy bolts fitting reamed holes. The wheel-shaft is $4\frac{1}{2}$ inches diameter, running in adjustable babbitt-lined bearings. A rope-wheel 4 feet in diameter is keyed on the shaft, and is grooved for fifteen $1\frac{1}{4}$-inch manilla "Stevedore" ropes, made with four strands and a core, worked in with plumbago in the process of making.

---
* F. E. Pritchard, *Elect. World*, April 16, 1892.
† See Trans. Am. Soc. C. E., vol. xxx., 1894.

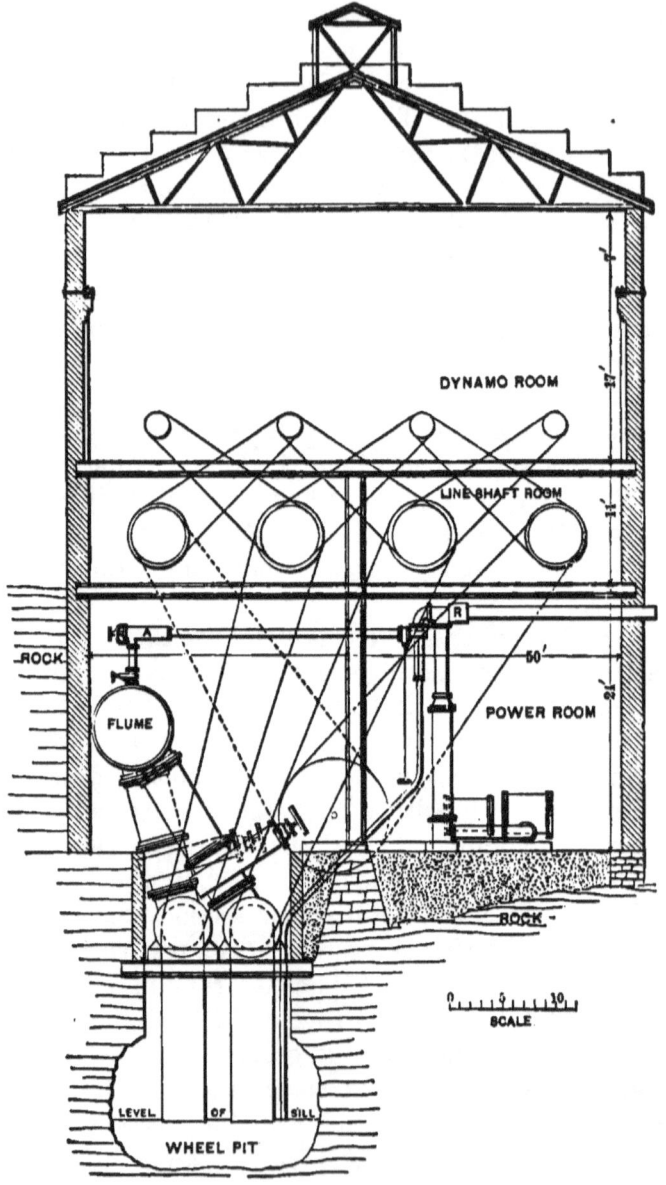

Fig. 31.—Section through Power Station.

From the 4-ft. wheel 15 ropes run to a rope-wheel on the line-shaft above, 76.8 inches in diameter, and grooved for sixteen 1¼-inch ropes. The ropes being endless, the idler-strand is passed over a 5-ft. single-grooved wheel, placed in a movable frame. The frame traverses in iron guides and maintains by its weight a constant tension on all the ropes. This is made adjustable for the amount of tension

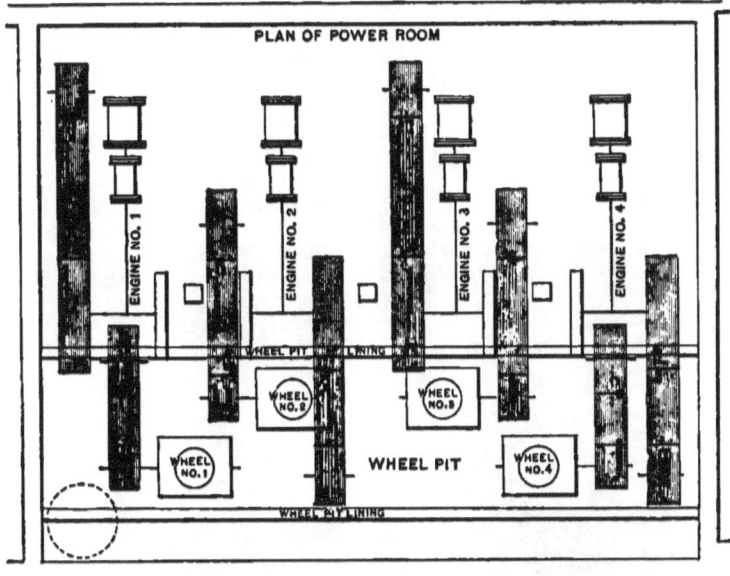

FIG. 32.—PLAN OF POWER STATION.

by the application of counter-weights to the frame. The speed of the line-shafts is 350 revolutions per minute, and the rope travel is 7037 feet per minute. The water-wheels are supplied from a steel flume 7 feet in diameter. From the horizontal portion of the flume a 4-ft. pipe leads down to each wheel, and has a geared 48-inch Chapman valve at the lower end, between pipe and penstock, as shown in

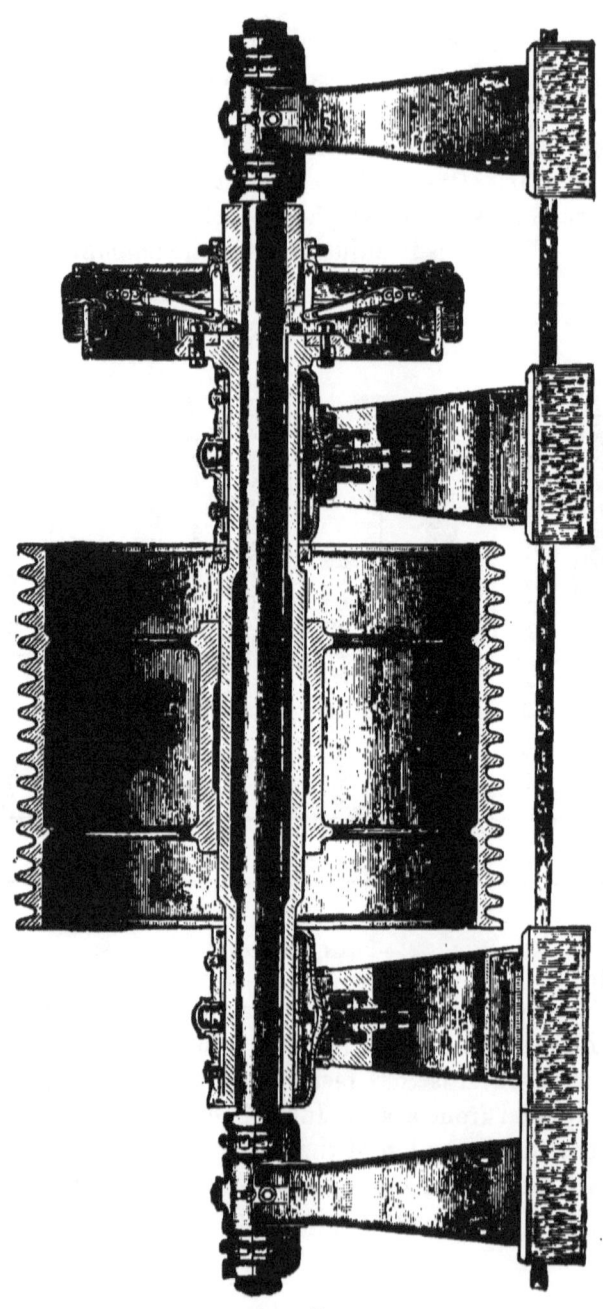

Fig. 33.—Method of Mounting Rope Pulley on Quill.

Fig. 31. These valves are fitted with a 12-inch by-pass, for the purpose of equalizing the pressure on both sides of the large valve in opening or closing.

A horizontal "Woodbury" compound condensing slide-valve engine, with extra heavy bed-plate, is set in the power-room at point marked in Fig. 32 "Engine No. 1." Steam-cylinders are placed with the large cylinder outside, so that pistons and rod may be easily removed. Cylinders are 19 inches and 31 by 24 inch stroke. At 167 revolutions, with a boiler-pressure of 110 pounds per square inch, vacuum 22 to 24 inches and cutting off at $\frac{5}{16}$ stroke, the engine is rated at 500 horse-power, and is guaranteed to produce a horse-power on an evaporation of 20 pounds of water per hour. The crank-shaft is a steel forging in one piece. Journals are $11\frac{1}{2}$ inches diameter by 21 inches long. Crank-pin $11\frac{1}{2}$ inches diameter by $8\frac{1}{4}$ inches long. The end carrying the rope-driving wheel has an outboard bearing. Governor balance-wheel is $8\frac{1}{2}$ feet diameter by 25-inch face. Rope-driving wheel is cast in halves 10 feet 6 inches diameter, and grooved for fifteen $1\frac{3}{4}$-inch ropes. These ropes lead to a 5-ft. rope-wheel on the line-shaft above, with same arrangement for tightener as is applied to the water-wheels. Rope speed of engine-drive is 5500 feet per minute.

The line-shafts are of hammered iron 5 inches in diameter, and arranged with heavy floor pedestals, fitted with self-adjusting, ring-oiled, babbitt-lined bearings. The rope-wheels are placed on heavy cast-iron quills, furnished with Hill friction-clutches of 500-horse power capacity each. By a series of jaw-clutches, pulleys, and belts any line-shaft can be operated from any water-wheel or engine, all the line-shafts making the same number of revolutions. Fig. 33 shows in detail the quill, clutch, and bearings.

CHAPTER IV.

THE use of ropes in connection with portable tools and travelling-cranes has long been established, and their convenience and adaptability to a wide range of work make them a necessity in many shops. The advent of the small electric motor in our machine-shops will, however, probably replace to a large extent all other forms of special transmission for portable tools, as it is already replacing countershafts and belting for machine-driving in many cases.

One of the greatest fields of usefulness for rope-driving is in the transmission of power to a moderate distance, under conditions which are unfavorable to the use of belts or shafting.

With rope-driving one is enabled at a comparatively small cost to transmit power in any direction to a building remotely situated from the source of power, which would otherwise require a long and expensive line-shaft or an independent engine or other motor. The facility with which it may be carried in any direction across rivers, canals, and streets, above or under ground, up hill and down, over houses and into buildings, is a feature very favorable to the further extension of rope transmission; but the rapid progress which has been made in the development of electrical transmission has limited the economical application of ropes to moderate distances. There are, however, certain limits between which the transmission of power by ropes is yet more efficient than by any other known method.

The employment of ropes for this purpose—i.e., transmission of power to a distance—is not a recent application.

The first method of transmission of power to any considerable distance was made in 1850 by C. F. Hirn at Logelbach, near Colmar, Alsace.* "The works consisted of a large number of buildings separated at some distance from one another, which were required to be changed into a weaving factory. As there was but one steam-engine on the works, the expense of transmitting power to the various buildings by ordinary shafting (the shortest length of which was 84 yards), or of erecting separate engines, would necessarily have been great; and the desire to obviate this expense resulted in the adoption of the telodynamic system.

The first plan adopted was the use of a band of steel 172 yards long, $\frac{1}{25}$ inch thick, and 2 inches broad. This was slung as an endless band over two wooden rollers or pulleys, 6 feet 6 inches diameter, which were placed 84 yards apart and made 120 revolutions per minute, giving a speed of 28 miles an hour in the band. In practice this plan was found to be open to two objections : the lightest wind agitated the band, and the pulley-guides tore it at the points of riveting, whilst the guides themselves were rapidly worn out. Notwithstanding these objections this plan rendered valuable service, and continued in operation for a year and a half, transmitting 12 horse-power to one hundred looms.

"The difficulties of the flat band suggested round wire ropes $\frac{1}{4}$ inch diameter ; these were accordingly substituted and were placed upon the same wooden pulleys, which, however, were first grooved to the depth of half an inch.†

This plan answered every expectation, and experience having fully sanctioned its use a second wire rope was soon put in operation, transmitting the power to a distance of

---

* Mr. H. M. Morrison, Proc. Inst. M. E. 1874, p. 57.

† Prof. Unwin, in his Howard lectures (see *Electrician*, Feb. 3, 1893), states that an English engineer, Mr. Tregoning, suggested the substitution of the wire rope.

about 770 feet. The two pulleys were each 9 feet 6 inches diameter, making 91½ revolutions per minute, and a steel rope ½ inch diameter was employed transmitting 50 h. p. at a speed of 31 miles per hour. In this instance it was found necessary to have supporting pulleys to prevent the rope from trailing upon the ground. These carrying pulleys were placed half-way between the transmitting pulleys, or 128 yards apart, and in the first instance they occasioned very great difficulties by the rapidity with which they were worn out in the groove. They were constructed successively of copper, wood, and polished cast iron, and were also faced with leather, horn, india-rubber, lignum-vitæ, and boxwood. All these failed, however; the facings were soon worn out, and when the groove was of metal or hard wood and did not itself wear it destroyed the rope. After repeated experiments a dovetailed groove was formed in the bottom of the pulley groove and filled with gutta-percha, (as shown in Fig. 65, page 186.)

"This turned out a perfect success, and carrying pulleys thus faced have an almost unlimited amount of durability." Fibrous ropes were used in the United States for long distance transmissions a few years later ; thus we find a communication in the *Scientific American** in which a correspondent from Winsted, Conn., speaks of several rope-drives in his vicinity, one of which had been in use since 1858. "It transmits the power for a manufactory, employing several circular saws, across the river, 225 feet distant, by a ⅝-inch rope running over two pulleys six feet in diameter, at a speed of 5600 per minute. The pulleys are sheltered, but the rope runs exposed in all kinds of weather, needing no attention except at times to be rubbed with grease having a very small amount of rosin mixed with it."

---

* Vol. III, 1861, p. 215.

ROPE-DRIVING. 65

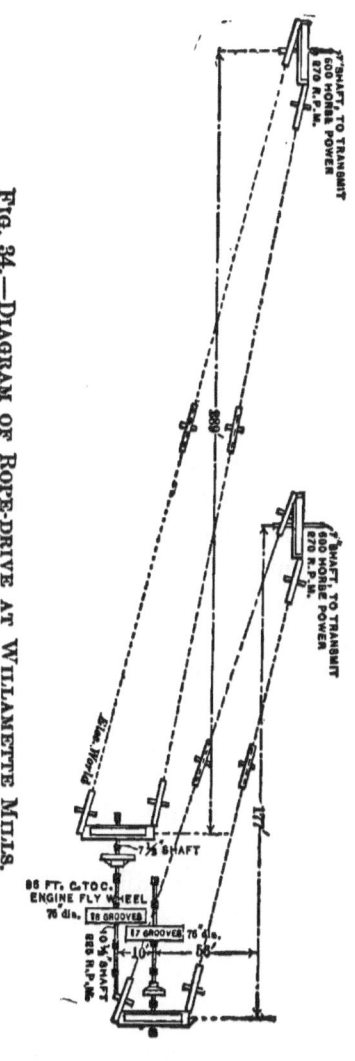

FIG. 34.—DIAGRAM OF ROPE-DRIVE AT WILLAMETTE MILLS.

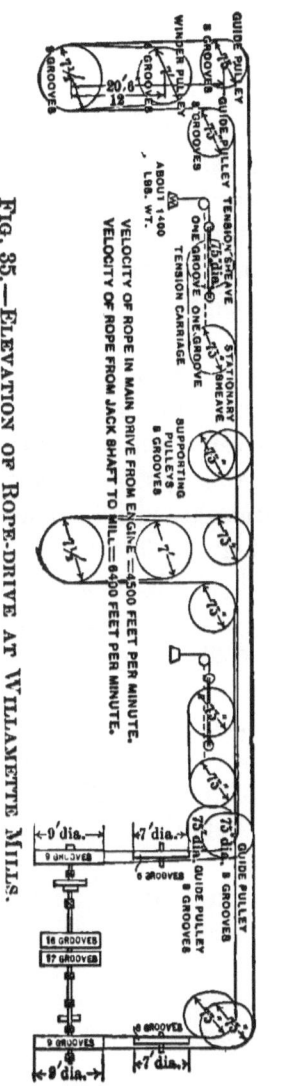

FIG. 35.—ELEVATION OF ROPE-DRIVE AT WILLAMETTE MILLS.

When power is taken from a water-wheel in locations where land is not available for buildings, the use of ropes as a means of transmitting the power from the wheel to the mill or factory forms a most economical arrangement if the drive is properly designed for the work. It is a great advantage in many other cases to have the power plant and the several buildings of a works isolated from each other; this is especially desirable in sawmills and wood-working establishments, where the risk from fire is greatly reduced by such an arrangement.

An example of rope-driving in which the conditions are particularly adapted to this form of transmission was erected a few years ago at Portland, Ore. The mill is built on piles situated in the Willamette River, while the engine and boiler room are upon the solid ground, some distance away. The engines are a pair of Wheelocks, with cylinders 32 by 60 inches, intended to be speeded to 70 revolutions per minute. The fly-wheel is 24 feet in diameter by 66 inches face; it is built up of ten segments grooved for thirty-three $1\frac{1}{4}$-inch manilla ropes, and fitted to a shaft 20 inches diameter; its weight is 40 tons.

Two ropes are taken from this wheel to jack-shafts 35 and 45 feet distant; the driven pulleys—one with 16 and the other with 17 wraps—are each 76 inches in diameter, and are keyed to 10-inch shafts. From the end of each jack-shaft a 600-h.-p. transmission is arranged and carried to the mill. Each shaft is fitted with a friction clutch to allow either of the mill transmissions to be thrown out if desired.

The general arrangement, showing location of driving and driven sheaves, is represented in Figs. 34 and 35, from which it is seen that power is delivered to two shafts 7 inches in diameter—in the one case at a diagonal distance of about 200 feet, and in the other at a distance of 185 feet from their respective drivers—both the driven shafts being at right angles with the jack-shaft. Each drive was designed to transmit 600 h. p. with a rope velocity of 7550 feet per minute, but it has since been found advantageous to reduce this speed to about 6000 feet per minute.

The arrangement of ropes in these transmissions is similar to that used in the Chicago City Railway Company, illustrated on page 53. In the present case only three wraps of $1\tfrac{1}{2}$-inch rope are used to convey the power from the jack-shaft to the mill, but in order to prevent slip and decrease the tension in the slack part of the rope the driving and driven pulleys have each nine grooves, six turns being carried around another pulley or winder, thereby increasing the arc of contact and, hence, the adhesion. In this plant the stress in the rope is very much greater than that ordinarily used.

It is evident that the whole strain must be borne by the three strands, as it is only the difference in tension of the tight and slack sides of the ropes that can be used to transmit power; since $\dfrac{PV}{33000}$ = h. p., we find the *difference* in tension, $P = \dfrac{33000 \times 600}{6000} = 3300$ pounds; and as there are three wraps, the difference in the stresses in the two portions of the rope will be 1100 pounds. As we shall find later the total stress will be greater than this, due both to the action of centrifugal force in the rope and to the force necessary for adhesion.

As the maximum working tension for a $1\tfrac{1}{2}$-inch rope is usually only about 450 pounds, it is evident that each wrap carries nearly three times its proper load, taking the wear and life of the rope into account. The engine has only developed about 700 horse-power as yet, so the total stress in each rope has been very much less than the above,—probably not more than 300 to 350 horse-power on each drive; but even with this reduced stress one rope was replaced inside of fifteen months.

For short distances shafting is often employed for such transmissions, but with this latter the friction of the journal-bearings is a very important consideration, and effectually debars its use for long-distance transmission.

This will be seen from the following considerations:

Let θ = distortion of shaft (circular measure) per unit length;
$\theta°$ = distortion in degrees;
l = unit length of shaft;
L = length of shaft in feet;
r = distance of outer fibres from axis = $\dfrac{d}{2}$;
d = diameter of shaft;
PR = twisting moment on the shaft;
N = revolutions of shaft per minute;
v = velocity of circumference of shaft = πdN;
G = modulus of torsion of the material
 = two fifths of the modulus of elasticity;
f = maximum torsional stress in the outer fibres
 $= \dfrac{16PR}{\pi d^3}$;
W = weight of shaft = 3.36 pounds per foot per square inch of section;
F = load due to friction.

Then
$$\theta = \frac{fl}{Gr} = \frac{32PRl}{\pi d^4 G} . \quad \ldots \ldots (1)$$

If the angle of torsion is given in degrees, then $\theta = \dfrac{\theta° \times 2\pi}{360}$; therefore the angular distortion per foot of length will be

ROPE-DRIVING. 69

$$\theta° = \frac{360}{2\pi} \times \frac{fL \times 12}{Gr} = \frac{360f}{\pi G} \times \frac{12L}{d}. \quad . \quad . \quad (2)$$

The working limit of the angle of torsion for steel shafting ought not to exceed 0.10 degree per foot in length of the shaft; that is, $\theta° = 0.10L$, and for wrought-iron shafting $\theta° = 0.075L$;* assuming the shaft to be of steel and substituting the corresponding value for $\theta°$ in (2), we obtain

$$0.10L \times \pi Gd = 360f \times 12L;$$

hence $f = 800d$ if we assume that $G = 11,000,000$ pounds.

Since the horse-power transmitted by the shaft equals $PR \times \frac{2\pi N}{33000}$, if we substitute the value of PR, $\left(= \frac{\pi d^3}{16} f\right)$, there is obtained h.p. $= \frac{\pi d^3}{16} f \frac{2\pi N}{33000}$; but the velocity at the circumference of the shaft is $v = \pi dN$, also $f = 800d$; hence

$$\text{h.p.} = 0.0095 d^2 v. \quad . \quad . \quad . \quad . \quad . \quad (3)$$

If the bearing is well worn and fitted to its shaft the resistance due to friction will probably lie between the limits $\frac{\pi}{2}\phi W$ and $\frac{4}{\pi}\phi W$,† or between $1.57\phi W$ and 1.28 ϕW, where ϕ is a coefficient, which in the present case we shall assume equal to 0.06.

Taking the lesser value, we shall have $F = \frac{4}{\pi}\phi W$, where F is the force at the circumference of shaft necessary to overcome the journal friction. If there are no pulleys on the shaft $W = \frac{\pi}{4} d^2 L \times 3.36$; the horse-power exerted to overcome the friction will then be

$$\text{h.p.}_0 = \frac{Fv}{33000} = \frac{4}{\pi} \phi \times \frac{\pi}{4} \frac{d^2 L \times 3.36v}{33000} = 0.0^5 6 d^2 Lv. \quad \text{Ex-}$$

---
* Reuleaux : Der Konstrukteur.
† Unwin.

pressed as a ratio, the percentage of power required to overcome friction will be

$$\frac{\text{h.p.}_o}{\text{h.p.}} = \frac{0.0^5 6 d^2 L v}{0.0095 d^3 v},$$

from which there is obtained

$$\text{h.p.}_o = 0.00063 \frac{L}{d} \times \text{h.p.} = \frac{L}{d} \times \frac{\text{h.p.}}{1585}. \quad . \quad . \quad (4)$$

That is, for a steel shaft whose diameter is one inch the horse-power required to overcome the friction in a length of 1585 feet will be equal to the total allowable transmitting capacity of the shaft.

For wrought iron $\theta^o = .075 L$ and h.p. $= .0075 d^3 v$, from which may be determined the value of the ratio

$$\frac{\text{h.p.}_o}{\text{h.p.}} = 0.0008 \frac{L}{d}, \quad \text{or} \quad \text{hp.}_o = \frac{L}{d} \times \frac{\text{h.p.}}{1250}. \quad . \quad . \quad (5)$$

The following tables, calculated from these formulæ, give the limits at which the power transmitted by a shaft is absorbed by the friction of the bearings, the assumption being that the factor of journal-friction ϕ equals 0.06, and that the allowable twist shall not exceed 0.10 degree per foot of length for steel, nor .075 degree for wrought iron. The shaft is supposed to be free from all pulleys and gears.

TABLE I.—LIMIT OF LENGTH FOR STEEL SHAFTING.
No pulleys on the line.

Diameter of shaft in inches.	Length in feet when total power is absorbed.	Length when efficiency = 50 per cent.	Length when efficiency = 75 per cent.
1	1,585	792	396
2	3,170	1,585	792
3	4,755	2,377	1,188
4	6,340	3,170	1,585
5	7,925	3,962	1,981

ROPE-DRIVING. 71

TABLE II.—LIMIT OF LENGTH FOR WROUGHT-IRON SHAFTING.
No pulleys on the line.

Diameter of shaft in inches.	Length in feet when total power is absorbed.	Length when efficiency = 50 per cent.	Length when efficiency = 75 per cent.
1	1,250	625	312
2	2,500	1,250	625
3	3,750	1,875	937
4	5,000	2,500	1,250
5	6,250	3,125	1,562

From the foregoing it will be seen that shafting is altogether unsuitable for conveying power any considerable distance, and as belting is not adapted to this work choice must be made of some other method.

For a mere dead pull, such as the alternate strokes needed to operate a pump, work is, and has long been, transmitted to very great distances; as by the long lines of draw-rods, ropes, or wires used in mining regions, quarries, and elsewhere, for transmitting the power of a water-wheel by means of a crank on its main axis, pulling during half its revolution, against a heavy weight at the end of the line, and thus storing up energy for the return stroke.

Wooden pump-rods were used in this manner about 1865 near Petroleum, W. Va. A large condensing engine was located in a central position, and the rods transmitted the power to a number of oil-wells, twenty-seven in all, situated at various distances and in different directions from the source of power. The greatest distance was about four miles.

Posts with crank-arms were used to change the direction of the pull. The rods were of hickory, connected end to end by means of iron straps.

The transmission of power from the famous 72-foot diameter overshot wheel at Laxey, on the Isle of Man, is by means of similarly connected trussed rods, which in this case are supported at regular intervals on small wheels running on iron ways. About 150 horse-power are transmitted in this way.

This method was adopted on a very large scale in the mines of Devonshire for the transmission of power from large overshot water-wheels to pumps fixed in the shaft of the mine at a considerable distance higher up the valley.

In one case* the water-wheel was 52 feet diameter, 12 feet breast, and its ordinary working speed was 5 revolutions per minute. The length of stroke given by the crank to the horizontal or "flat" rods was 8 feet; the rods were $3\frac{1}{2}$-inch round iron, and were carried on cast-iron pulleys.

At Devon Great Consols, near Tavistock, there are alto-

FIG. 36.—ROPE-DRIVES WITH BENT CRANKS AT 120 DEGREES.

gether very nearly three miles of 3-inch wrought-iron rods, carried on bobs, pulleys, and stands, whereby power for pumping and winding is conveyed along the surface to different parts of these extensive mines, from 11 large water-wheels ranging up to 50 feet in diameter, to which the water is brought along eight miles of leats 18 feet in width.

Rods and wire ropes have also been used to transmit rotary motion to a considerable distance in a similar manner by placing the cranks at 120 degrees, as shown in Fig. 36.

It is evident that the distance of transmission by this contrivance will be subject to the sag of the ropes, unless

---

* "The Old Wheal Friendship Mine, near Marytavy." Proc. Inst. M. E. 1881, p. 100.

intermediate shafts are employed. The motion must also be comparatively slow, owing to the severe strains which would be thrown upon the bearings and pins by the surging and swaying of the ropes during the rapid changes of motion to which they would be liable. In order, then, to transmit much power, heavy rods or large ropes would be necessary, and under these conditions economical transmission would be limited to short distances.

Among the various means in use at the present time for conveying power to a distance we find steam, water, gas, compressed air, electricity, and rope systems. Each of these has its own applications and advantages, but it must be borne in mind that with the exception of rope transmission, of which numerous examples have already been given, all other forms usually require a generator at the one end and a motor, with separate attendants, at the other.

Other things being equal, the relative merit of various methods of transmitting power will be indicated by the cost of transmitting a certain amount of power to any given point, as compared with the cost of this power at the generating station, while their absolute merit will be shown by comparing the cost of the transmitted power at the receiving station with the cost of producing the required power directly at this point.* Such determinations are materially affected by variations in the amounts of power and in the distance of transmission; the other principal factors to be considered being the efficiency of the system, the number of working hours per annum, the price of 1 h. p. per hour at the generating and receiving stations, and the convenience and applicability of the system to each special case.

The efficiency of any system of transmitting power is expressed by the ratio of the power obtained at the receiv-

---

* Stahl, "Wire-rope Transmission."

ing station to the power given out at the generating station. In all systems losses of power of greater or less magnitude occur, and the most efficient system is that in which those losses are reduced to a minimum. We shall not attempt here to lay down rules governing the choice of any particular method, for the requirements and conditions are so varied that every individual case must be decided upon by the engineer separately with a knowledge of all the facts before him. Our present object is to ascertain the principles governing the use of ropes, and to determine those conditions best suited to their economic working.

CHAPTER V.

THE subject of rope-driving may properly be placed under two heads, according to the nature of the material composing the ropes, whether metallic or non-metallic. With few exceptions metallic or wire ropes are used almost exclusively on long-distance or telo-dynamic transmission, while non-metallic ropes are employed for intermediate and comparatively short drives, the consideration of which constitutes the present subject-matter.*

Among the materials employed in this method of power transmission we find special forms of leather belting used as ropes working in V grooves; fibrous ropes, including flax, hemp, cotton, and manilla, are, however, chiefly employed.

Rawhide ropes, which are made from $\frac{3}{8}$ inch to 2 inches in diameter, are used to a limited extent. Where the stress in the rope is not great and the accompanying slip is small, rawhide works very well, and will last from three to six, and in some cases ten, years. Under ordinary circumstances it is not necessary to use any dressing, as sufficient lubrication is furnished by the rope itself; if the rope slips in its groove the leather will be burned, and lose its flexibility, and also its adhesive qualities, to a certain extent. A rawhide rope has very little tendency to rotate

---

* Concerning wire-rope transmission the reader is referred to the following:

"Wire-rope Transmission" (A. W. Stahl); "Elektrischen Kraftübertragung" (A. Beringer); "Drahtseiltriebs" (D H. Ziegler); "Constructeur" (F. Reuleaux); "Machine Design" and "Central Stations" (W. C. Unwin). Also trade pamphlets published by W. A. Roebling & Sons; Cooper, Hewitt & Co.; and others.

on its axis; for this reason the wear is not always uniform, and with a heavy tension it is liable to take the set of the groove in which it runs. This is rather an advantage for a straight drive, where the rope always runs in the same direction; but in those cases where a rope is led on to the pulleys at an angle this will be a disadvantage, as under such conditions the rope often slips, and wear is excessive. Where the rope is subject to wet or dampness, rawhide is an excellent material to use, as it is very little affected by dampness. The cost of rawhide rope will average about six times that of a good quality of manilla transmission-rope, and although it is to be preferred in certain cases, its greater cost will limit its application.

Round-leather ropes, formed by twisting narrow strips of leather into a continuous spiral, are used for light driving, and are very desirable for some classes of work.

Solid round-leather ropes, made from several thicknesses of belting cemented together and secured with screwed wire forced into the leather, are made in various sizes up to $2\frac{1}{2}$ inches in diameter, but sizes larger than $\frac{3}{4}$ or 1 inch in diameter are seldom used.

Steel ropes with leather washers closely threaded on have been tried with considerable success, but the expense of such a rope would necessarily limit its application.

Other special forms of leather belting used as ropes are found in the various modifications of the square and angular belts which have been used for a number of years for both light and heavy drives.

Leather ropes as large as $1\frac{3}{4}$ inches square, made up of layers of leather cemented together so that the whole is uniform and continuous, have been used to replace quarter-turn flat belts, and also for main driving.

These run in V grooves so that the adhesion is greatly in excess of that produced by a flat belt on a smooth pulley under the same tension. In the same way triangular belts built up from various thicknesses of leather possess the advantages characteristic to all forms of rope-driving which use a V groove, viz., greater adhesion for a given tension, and the facility with which such transmitters lend themselves to the communication of power between shafts at an angle with each other.

It is evident that several of these leather-rope belts may be used side by side in a manner similar to the various applications of fibrous ropes. Such ropes have proven satisfactory in those cases where the pulleys are of approximately the same diameter; but on pulleys whose diameters vary considerably each portion of the leather rope in contact with the driver tends to rotate the follower at a different velocity, necessarily producing slip and wear, to an extent depending upon the ratio of the diameters employed.

In England manilla is now being used very largely, but cotton ropes were formerly preferred to the exclusion of all others for all kinds of driving; but the most probable cause of this was not that cotton was the best or most economical material for the purpose, but that rope-driving is most common at cotton factories, and cotton ropes were made in the locality by men who were familiar with the local product, and had been for years making spindle and rim bands of small size. When the demand for large sizes arose these rope-makers applied themselves to the newer industry, and shut out other materials.*

In the mills of Dundee and vicinity, and in the North of Ireland, where flax and hemp are worked, we find ropes of hemp, a local product, used entirely.

In many cases ropes of cotton are to be preferred, as they are generally softer and more pliable than the ordinary manilla ropes, thus allowing smaller pulleys to be used

---

* W. H. Booth, *Am. Machinist*, January, 1891.

with less injury to the fibres. In fact, cotton ropes of small diameter have been used for years in cotton machinery bandings over pulleys, and under conditions which would wear out a manilla rope in one third the time. There is also an advantage in that there is less internal chafing and wear when the rope is bent over a pulley, on account of the smoothness of the fibres and the great elasticity of the yarns.

The cotton fibre is not, as it appears to the eye, a solid cylindrical, gossamerlike hair, but when fully ripe is shown under the microscope as a flattened hollow ribbon or collapsed cylindric tube twisted several times throughout its length, as shown in Fig. 37;* it is of equal size for about three fourths of its length, and it then gradually tapers to a point. This point is a section almost perfectly cylindric, and, unlike the rest of the fibre, often composed of solid matter. Covering the outside membrane of the fibre is an oleaginous coating generally known as cotton-wax. This wax amounts to about two per cent of the fibre, and in the spinning of the material it requires to be reduced to a certain point of liquefaction by the heated temperature of the room before it can be made to work properly without lapping on the drawing rollers.

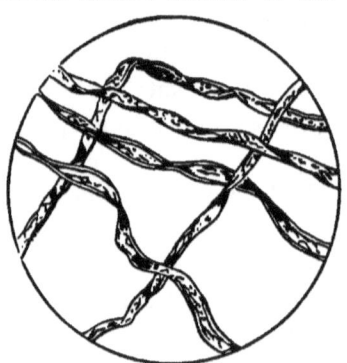

FIG. 37.—COTTON FIBRE, ORLEANS VARIETY (*Gossypium Hirsutum*).

These fibres vary in size from 0.00084 inch mean diameter, and about ½ inch long to 0.000635 inch mean diameter and 2⅛ inches in length, depending upon the variety of the cotton; but for a given variety the differ-

---

* See "The Cotton Fibre," by Hugh Monie, Jr. Published by Heywood & Son, Manchester, Eng.

ence is very small: thus in the Sea Island cotton the maximum length of fibre is 2 inches, while the minimum is $1\frac{3}{4}$ inches; in the same way in the Orleans variety shown in Fig. 37 the maximum length is $1\frac{1}{8}$ inches and the minimum $\frac{1}{1}\frac{5}{6}$ inch.

As a rule, those cottons which have the longest fibres are also the smallest in diameter: they possess the natural twist in a more perfect and highly developed form, and are much stronger and more elastic.

In all good commercial fibres of cotton there is necessarily (1) a very small percentage of solidified oleaginous matter distributed over the internal surface of the fibre deposited when the vital fluids were in active circulation; and (2) a certain percentage of moisture known as water of hydration.

These together with the twisted structure impart to the fibres that suppleness, tenacity, and elasticity without which they would be almost useless for manufacturing purposes.

The cotton fibre is thus naturally well adapted to the work of being twisted into yarns; the presence of the natural convolutions and comparative smoothness of the surface of the unit filament permits considerable elongation, and the wax on its surface serves as a natural lubricant and prevents the fibres from becoming brittle.

Thus it will be seen that ropes made from fibres possessing these characteristics are particularly well adapted to the transmission of power in which the rope is constantly undergoing a varying strain and is subjected to much flexion.

The strength of cotton ropes is, however, relatively small when compared with other fibrous ropes, and although the weight is about one fifth less than manilla, for equal diameters, the actual first cost is from fifty to seventy five per cent greater than for the latter.

Nystrom gives the breaking strength of three-strand

cotton ropes at less than one tenth that of similar manilla rope, but this is apparently too low for a good quality of transmission rope.

Tests made at Watertown on a number of Lambeth ropes varying in size from 1 inch to $2\frac{1}{8}$ inches in diameter indicate that the breaking strength is equal to about $4000d^2$ pounds, while the extension varies from twenty to twenty-five per cent, corresponding to a reduction in diameter of about fifteen per cent. The weight of these ropes is very closely $0.26d^2$ pounds per foot of length.

A series of tests carried out by Kircaldy* on cotton ropes ranging in size from $1\frac{3}{8}$ inch to $2\frac{3}{16}$ inches in diameter give the breaking strength as $3700d^2$ pounds for minimum value and $5800d^2$ pounds as a maximum. The weight per foot of these ropes varied from $0.25d^2$ to $0.29d^2$; the extension under a stress of about 85 per cent of the breaking load varied from 17 to 27 per cent.

Reduced to a common basis in which the strength is made proportional to the weight, and averaging the results, we find that the breaking strength may be represented by $4600d^2$ pounds.

The data on cotton ropes are too meagre to determine whether their strength decreases as the diameter increases, but this is probably the case.

Rouleaux gives 7500 pounds per square inch of section for cotton transmission-ropes, which agrees very closely with the above values.

From the formula, breaking strength, $S = 4600d^2$ pounds the values given in Table III have been calculated, and may be considered as representing approximately the strength of cotton transmission-ropes of good quality.

The working strength of cotton transmission-rope may be taken higher, in proportion to its ultimate strength,

---

* See also Kent's "Mechanical Engineer's Pocket Book."

TABLE III.—STRENGTH OF COTTON TRANSMISSION-ROPES.

Diameter of Rope in inches, d	Ultimate Breaking Strength, $S = 4600d^2$
$\tfrac{1}{2}$	1,150
$\tfrac{5}{8}$	1,800
$\tfrac{3}{4}$	2,600
$\tfrac{7}{8}$	3,500
1	4,600
$1\tfrac{1}{8}$	7,200
$1\tfrac{1}{4}$	10,400
$1\tfrac{1}{2}$	14,000
2	18,400

than is used with manilla, for the latter is weakened by the grease with which it is lubricated; and, moreover, a larger factor must be allowed for wear on account of the character of the manilla fibre, which breaks more easily under bending strains.

As compared with manilla, then, the advantages of cotton ropes of the same diameter are: Greater flexibility, greater elasticity, less internal wear and loss of power due to bending the fibres, and the use of smaller pulleys for a given diameter of rope. Its disadvantages are: Greater first cost, lesser strength, and, possibly, a greater loss of power due to pulling the ungreased rope out of the groove —in any case this is usually small with speeds over 2000 feet per minute.

As we have already noted, manilla rope is used very extensively for transmission purposes, but its application has not always met with that success which would follow a more thorough knowledge of its requirements. Inefficient rope-drives are erected and run for a few months, or perhaps only days, and are replaced with larger ropes if the sheaves will permit, or, as in many cases, the ropes give way to leather belting, and henceforth rope-driving is condemned. The true cause is not so much the inefficiency of the ropes as it is the lack of knowledge governing their use and application; in order to obtain a proper conception of this a study of the structure of the rope will be found advantageous. Manilla, or, more properly, manilla hemp (abaca) rope, is made from the fibres of the *Musa textilis*, a plant closely allied to the banana, growing near Manilla, in the Philippine Islands. The fibres are a part of the outer covering of the leaf-stalk, which attains a length sometimes as great as 15 feet. To obtain fibres of suitable size for manufacturing rope the leaf stalks are subdivided, and in the process of segregation the fibre assumes an appearance somewhat similar to that produced in splitting a piece of wood—it is rough and uneven, and more or less splintery throughout its length. These fibres, although in themselves not very large, are composed of very fine and much elongated bast-cells, which overlap each other as shown in Fig. 38. The cells are irregular in outline and vary considerably in size. The length of the cells is about one fourth of an inch. A series of tests on manilla fibre carried out by Dr. Stanley M. Coulter of Purdue University, shows that the cells are not, as commonly supposed, held together by an intercellular tissue or mucilaginous substance. A cross-section of a portion of the fibre, (*a*), Fig. 39, enlarged 450 times, shows that there are no intercellular spaces; and various reactions to determine the presence of mucilage or other vegetable glue revealed no traces of its presence.

Fig. 38.
Cells of Manilla Fibre, enlarged.

The characteristic roughness possessed by the manilla fibre is due entirely to mechanical causes, such as, for instance, the laceration of a cell in the separation from the leaf-stalk, or the subsequent opening out of the ends of the cells.

The contour of the perimeter is also rough, as noted in the figure, 39b, as it retains the form impressed upon it by the contiguous cells when in the plant.

These fibres have great strength in the direction of their length, but are weak transversely;* when made into rope they are compelled, in bending over the sheave, to slide on

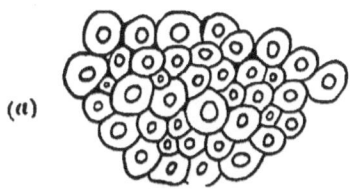
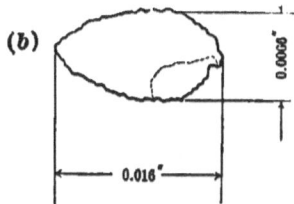

FIG. 39a. — MANILLA FIBRE, MAGNIFIED 450 TIMES.

FIG. 39b. — CROSS-SECTION OF MANILLA FIBRE.

each other while under pressure from the load. This causes the internal chafing and grinding which, if not prevented, soon wears out a rope when subjected to bending strain.

In addition to the action of the fibres upon each other, the strands and the yarns of which the strands are composed also slide a small distance upon each other, causing friction, and hence internal wear.

By opening out an old dry rope which has been used over a sheave, a fine powder will be disclosed, showing that where the rope was bent over the sheaves the strands, in sliding on each other, ground some of the fibres to powder. Another reason for this is that the fibres in an old rope become brittle and weak when dry, so that the constant

---

* The tensile strength of manilla fibres will average over 30,000 pounds per square inch of section.

flexure to which they are subjected rapidly disintegrates the cell congeries.

Aside from the external wear which a rope suffers from contact with its sheave (part of which is the differential driving effect), these two are the principal causes of rapid wear in a rope-drive, to remedy which we must in the first case lubricate the fibres, and in the second, prevent undue flexure of the rope. How this is effected in practice will be seen presently.

CHAPTER VI.

In manufacturing manilla rope the fibres are first spun into a yarn, this yarn being twisted in a direction called right-hand. From 20 to 80 of these yarns,* depending on the size of the rope, are then put together and twisted in the opposite direction, or left-hand, into a strand. Three of these strands for a 3-strand, or four for a 4-strand, rope are then twisted together, the twist being again in the right-hand direction. It will be noticed that when the strand is twisted it untwists each of the threads, and when the three strands are twisted together into rope it untwists the strands, but again twists up the threads. It is this opposite twist that keeps the rope in its proper form. When a weight is hung on the end of a rope the tendency is for the rope to untwist and become longer. In untwisting the rope it will twist the threads up, and the weight will revolve until the strain of the untwisting strands just equals the strain of the threads being twisted tighter. In making a rope it is impossible to make these strains exactly balance each other, and it is this fact that makes it necessary to take out the turns in a new rope—that is, untwist it when it is put at work.

---

* A three-strand rope one inch in diameter is the key to the sizing of the yarns. Yarns of 20s are of such a size as to require 20 to fill a tube half an inch in diameter, or to make one strand of an inch rope; 26s requires 26 to fill the same size tube; and so on. The size of cotton yarns, on the other hand, depends upon the number of hanks per pound; thus 20's, or number 29 cotton, requires 20 hanks (340 yards each) of this size to weigh one pound.

The fibres of manilla which are thus twisted into ropes will average over 6 feet in length, varying from 3½ to 12 feet. If they were long enough, the most advantageous method of using them would be to lay the fibres side by side, and secure them at the two ends; each fibre would then bear its own share of the strain, and the strength of the bundle would be that of the sum of the strengths of the separate fibres. As a long rope could not be formed in this way, the fibres are secured by twisting so as to produce sufficient compression to prevent them from moving upon each other when a strain is applied; but in attaining this amount of compression their strength is greatly reduced; this very compression acts as a constant weight on the fibre, and must be deducted therefrom before the available strength can be applied.

The weakening effect produced by twisting varies considerably among the fibres of the same rope according to their distance from the centre or heart of the bundle. If a certain amount of twist be given to a bundle of fibres, the outer ones will be strained more, and will act with less useful effect than those on the inside, which will have to bear the greater part of the strain while the rope is being used. It will therefore be evident that if the fibres were twisted at once into a thick rope the outer fibres would be so much strained as to be of little or no use in contributing to the strength of the rope, but by making the rope as we have indicated by first twisting into yarns, then into strands, and finally combining these into a rope, the strain is more equalized, and the important properties of length and strength are secured without too great a sacrifice of the strength of the individual fibres.

The degree of twist in the rope may be determined by constructing a right-angled triangle, the base of which is the circumference, and the height the length of one turn of the strand measured parallel to the axis. The difference between this height and the hypothenuse is the quantity by which the rope is twisted. The ropemaker's ordinary rule for a three strand rope is to have one turn to as many inches as are contained in the circumference of the rope; but the degree of twist is variable and more or less dependent upon the judgment of the maker.

Experiments by Réaumur to determine the effect of twist upon a rope showed that a small well-made hemp cord broke in different places with a mean weight of 65 pounds; while the three strands of which it was composed bore $29\frac{1}{2}$, $33\frac{1}{2}$, and 35 pounds respectively, so that the total absolute strength of the strands was 98 pounds, although the average real strength was only 65 pounds, thus showing a loss of 33 per cent.

More recently the test of a small rope showed an average strength of 4550 pounds, while the aggregate strength of its 72 yarns was 6480 pounds—each yarn bearing about 90 pounds; thus there is a loss of 1930 pounds, or about 30 per cent.

Ropes made of the same hemp and the same weight per foot, but twisted respectively to two thirds, three fourths, and four fifths of the lengths of their component yarns, supported the following weights in two experiments made by Duhamel:*

	Pounds.	Pounds.
Two thirds	4,098	4,250
Three fourths	4,850	6,753
Four fifths	6,205	7,397

The results of these experiments led Duhamel to make ropes without twist by placing the yarns together and wrapping them round to keep them together. The rope had great strength and pliability, but not much durability

---

* Tomlinson, vol. VII. p. 574.

on account of the outer covering wearing away or opening when bent, thus admitting moisture to the interior, which rotted the yarns.

In general, the greater the twist the more hard and rigid the rope is, and the better it will keep its form; but it is not as strong, weight for weight, as the more loosely twisted rope; moreover, the hard-twisted rope is more rigid than the other, and is not as suitable for transmission purposes, owing to the rapid wear which constant flexure produces.

A very excellent transmission-rope, known as the "Lambeth," is made in a somewhat similar manner to that just described, but it is not open to the same objections. In this case the rope, which is of cotton, is made of four strands twisted together in the usual manner, but the strands are themselves composed of a bundle of fine yarns and have scarcely an appreciable twist. Each bundle, comprising many hundred yarns, is wound spirally with smaller bundles of about 100 yarns each. In this way the outside of the rope acts as a shield or covering to the cores which do the work. By this means a certain amount of the natural elasticity of the cotton is retained, and its pliability is much greater than in ordinary hawser-laid ropes.

Lubrication of transmission-ropes is provided for in various ways. Frequently the rope is laid up dry and a coating or dressing is given to the exterior, which is supposed to penetrate to the interior and lubricate the fibres. With some dressings this may occur, but with others the effect is merely local; the interior of the rope remains dry, and much bending soon wears it out. With cotton ropes, as we have already noted, the internal chafing and wear is very much reduced, and for this reason cotton ropes are laid up dry and are not usually lubricated; they are, however, generally coated with some form of dressing to prevent the fibre from rising or the rope fraying, and to protect it from an undue amount of moisture when exposed to the weather; it also assists in retaining the natural moisture in the fibres, without which they would become brittle and weak.

According to the nature of the dressing, the interior may or may not be affected by the outer coating. Beeswax and black lead, with a little tallow, forms an excellent moisture-proof covering for ropes; it fills in the spaces between the strands, and the rope soon assumes a perfectly round and smooth appearance, like a bar of iron; with this coating the interior retains its natural condition, and should therefore not be used where a lubricant is desired. Pine tar is much used on cotton ropes for the same purpose. Mixtures of tallow and black lead; molasses and black lead; equal parts of resin or beeswax, with black lead, tallow, and molasses melted together, and applied hot; and various other compounds are in use for this purpose. Tallow, lard, and other greases are used separately, and are fairly satisfactory as a lubricant. When the rope runs out-of-doors a water-proof coating is necessary to preserve it from decay, and for this purpose, if it is also desired to lubricate the rope at the same time, there is probably nothing better than tallow and black lead, or graphite, provided the rope is not twisted too hard, which would prevent the dope from penetrating. In certain drives that are subjected to hard service and are more or less exposed to the weather boiled linseed-oil is used very successfully. According to Mr. A. D. Pentz,* the rope is treated every two weeks to about two quarts of the oil, dripped one drop at a time upon one of the sheaves, which is uncovered on top; the rope runs on the bottom of the sheave and slowly

---

* *Eng. Magazine*, Nov. 1893, p. 256.

absorbs the oil. Although the rope is very materially weakened by this process, yet the greater freedom of the fibres permits a heavier working strain to be carried, for it is the relative wear of the fibres that determines the life of the rope. A manilla rope with the fibres properly lubricated will, under the same conditions, outlast from two to four similar dry-laid ropes which are allowed to run dry. Manilla transmission-ropes are generally laid up in tallow, paraffine, soapstone, or a mixture similar to the above preparations.

A superior manilla transmission-rope is that known as the "Stevedore," or black rope, which is made with both three and four strands, the latter being laid up about a central core. The yarns of this rope are each coated with a mixture of graphite and tallow, so that when twisted into strands the coating lodges in the hollows and uneven places among the fibres, and thoroughly lubricates the strands and individual fibres composing the rope, which is thus made practically as nearly water-proof as possible. After it has been in use a short time its appearance is that of a black rod of iron, smooth and round, similar to the beeswax-coated rope previously mentioned, but perfectly flexible. A manilla rope thus made will last from three to eight years if not overstrained; when running indoors under favorable conditions the latter limit may be attained, but when exposed to the weather, or when working under less favorable conditions, its life will be shortened.

Hemp ropes intended for outdoor service are sometimes treated by passing the yarns through boiling-hot tar, suitable machinery being used to regulate the amount of tar retained in the yarns so that the fibres may be coated over and thus preserved from decay. Tarring protects rope from injury by exposure to rain and immersion in water, but it makes its fibre rigid and impairs its strength; for this reason it is unsuitable for ordinary transmission purposes.

It has been shown by experiment:*

. 1. That white or untarred rope in continual service is one third more durable than tarred. 2. That it retains its strength much longer when kept in stock. 3. That it resists the ordinary injuries of the weather one fourth longer.

With the exception of the outside yarns of large hawsers, manilla ropes are not tarred.

The breaking strength of a rope depends both upon the quality of the material and the degree of twist given to the strands; for a loosely twisted rope of a given diameter the strength is less than that in a hard twist of the same diameter, but compared weight for weight, the rope with the lesser degree of twist is the stronger.

In discussing the strength of ropes, which formerly was always given in terms of the circumference, there is a lack of uniformity among writers in the relation between the diameter and area of a rope. The circumference, as measured by a tape, depends upon the number of strands in the rope and their compression upon one another. If the strands still retained their circular section when twisted into a rope, the circumference of a 3-strand rope would be 2.86 times the diameter of the circumscribing circle, as given by Nystrom; if the strands completely fitted the circle its measure would be π times the diameter as given by Unwin and others.

As neither of these conditions obtains in practice the true value lies between 2.86 and 3.14, and we shall assume 3 as the most suitable factor.

In the same way the area of the cross-section of the rope is variously estimated as that of the area of the circular strands, or as the area of the full diameter of the rope.

---

*Duhamel: Traité de la fabrique des manœuvres pour les vaisseaux.

By assuming the rope to be made of three strands, the cross-section of each of which is a circle, the area of the rope would evidently be $3\left(\dfrac{\pi}{4}\delta^2\right)$, where δ is the diameter of each strand. If d represents the diameter of rope, i.e., the diameter of circumscribing circle, the area in terms of d will become

$$A = 3\left(\dfrac{\pi}{4}\delta^2\right) = \dfrac{\pi d^2}{6.16} = 0.52 d^2.$$

The section of the strands, taken at right angles to the axis of the rope, is, however, not a circle, as can be seen from Fig. 40. The degree of twist given to the strands,

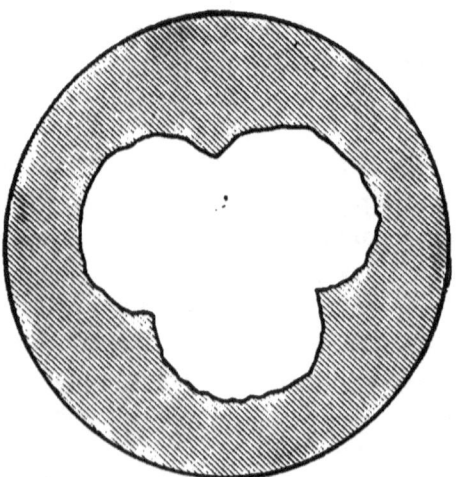

FIG. 40.—ACTUAL SECTION OF ROPE.

and the compression of the latter upon one another, will evidently affect the area of the section; for the longer the spiral the more nearly will the cross-section of each strand approach a circle. It is obvious that the true value must lie between $0.52 d^2$ and $0.7854 d^2$. In determining this area the writer made a number of plaster casts at different points of several 3-strand manilla ropes, varying in size from ¾ inch to 1¾ inches diameter. With these casts as dies, which were covered with printer's ink, impressions were made and the area obtained by using a planimeter. The area at several sections was found to be practically constant for each rope, and varied between $0.61d^2$ and $0.65d^2$—the mean value being $0.63d^2$, from which we obtain the ratio

$$\frac{\text{Area of section of rope}}{\text{Area of circumscribing circle}} = \frac{0.636d^2}{0.7854d^2} = 0.81;$$

that is, the actual area of a 3-strand rope equals eight tenths of the area of the circumscribing circle.

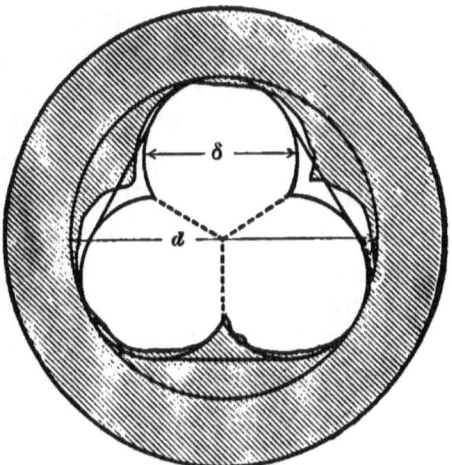

FIG 41 —CROSS-SECTION OF ROPE.

A print obtained with a plaster cast, showing the distortion of the strands from a true circle, is reproduced in Fig. 40. Fig. 41 is a similar print, showing the relation between the actual area of the cross-section and the area included between circular strands and the tangents joining them. It will be noted that the excess of area outside of the lines drawn tangent to the inner circles is but slightly greater than that between the tangents and the inner circles.

An inspection and tabulation of the results of numerous tests on manilla rope * shows that the strength per square inch increases as the diameter of the rope decreases. Formulas for the strength of rope based upon the circumference and a constant multiplier—as, for instance, $S = 800c^2$, where S is the breaking strength and c the circumference—must be regarded as giving only an average value for diameters approximating those experimented upon. As the strength of good manilla rope varies from 10,000 pounds per square inch for a 2-inch rope to over 12,000 pounds for a half-inch rope, it can be seen that a more accurate determination may be made if a variable is used in the formula. With the above assumption of area ratios, obtained by trial, the following expression has been deduced, which will give a very fair value of the breaking strength for new manilla ropes: $S_1 = 100d^2x$, in which x is a variable depending upon the diameter of rope; for manilla rope we may assume the empirical value $x = 81 - 9d$, where d is the diameter in inches.

The ultimate strength per square inch of actual section will then be

$$S = \frac{S_1}{0.8\frac{\pi}{4}d^2}.$$

From these formulas Table IV has been computed:

The above values are for new manilla ropes made of selected stock; ropes that are greasy or wet will be reduced

---

* Major Parker's report of tests made at Watertown Arsenal, 1885. Ex. Doc., No. 36. Experiments of M. Doboul in "Bulletin de la Société d'Encouragement des Arts," Paris, 1888. Riehle Bros.' laboratory tests; and others.

ROPE-DRIVING.

TABLE IV.—STRENGTH OF MANILLA TRANSMISSION-ROPES.

Dia. of rope = d.	Breaking Strength, $S_1 = 100 d^2 x$.	Strength per square inch, S.
½	1,900	12,200
⅝	2,900	11,950
¾	4,100	11,750
⅞	5,500	11,525
1	7,100	11,350
1⅛	8,800	11,125
1¼	10,900	10,975
1⅜	15,000	10,625
1½	19,800	10,300
2	25,100	10,000

in strength from 20 to 30 per cent, but when ropes are used for transmission of power the lubrication of the fibres is of more importance than the actual breaking strength of the rope, as in any case the apparent working strain, as calculated from the power transmitted and the speed of the rope, should not be taken greater than 5 per cent of the ultimate strength; in many cases not more than 2 per cent is used.

If we wished to obtain the working strength in any given case we would usually divide the breaking strength by an assumed apparent factor of safety, but for a flying rope, in addition to this, it is necessary to provide a factor of wear; moreover, the actual strains are generally much greater than the normal as calculated from the power transmitted, due to vibrations in running, imperfect tension mechanisms, defects in construction, and other causes. As the strength at the splice is only 70 to 75 per cent of the strength of the rope, the actual margin for wear and unknown strains is not as large as would at first seem apparent. In the first place, the strength of a lubricated rope is weakened about 25 per cent by the grease, then 25 per cent more must be deducted for the strength of the splice; this leaves only about 50 per cent of the original strength of the rope which is

TABLE V.—EXECUTED ROPE TRANSMISSIONS.

H.P. of Engine.	Diam. of Rope.	No. of Ropes or Wraps.	Velocity of Ropes. Ft. per Min. $= V$.	Driving Force $= P$.	Probable Ratio Max. Tension to Driving Force $= \frac{T_1}{P}$.*	Corresponding Value of Max. Tension $= T_1$.	Assumed Allowable Strain from $T_1 = 200d^2$.	Material of Rope.	REMARKS.
2000	$1\frac{3}{4}$	30	5089	367	1.66	608	612	Cotton	English.
1350	$1\frac{3}{4}$	27	5186	318	2.3	730	612	Manilla	Water-wheel.
1300	$1\frac{3}{4}$	32	5338	251	2.35	590	612	....	English.
1200	$1\frac{5}{8}$	36	4398	250	2.12	530	528	....	English.
1200	$1\frac{3}{4}$	27	5277	278	2.35	680	612	Manilla	In use 3 years; good condition.
1200	$1\frac{3}{4}$	20	5100	390	2.0	780	612	"	
1100	$1\frac{5}{8}$	35	4490	230	2.12	485	528	Cotton	English.
1100	$1\frac{3}{4}$	26	4900	275	2.25	620	612	Manilla	
1062	$1\frac{3}{4}$	27	5086	255	2.25	575	612	....	English.
1000	$1\frac{3}{4}$	23	5227	282	2.35	660	612	Manilla	
800	$3\frac{1}{2}$	24	1230	890	1.70	1,520	2,450	Cotton	Used 5 years without resplicing.
750	2.36	30	3000	275	1.86	510	1,114	Hemp	German.
450	$1\frac{1}{4}$	14	4850	220	1.66	365	312	Cotton	In use 6 years; good condition.
350	$1\frac{3}{4}$	12	5166	186	1.66	308	612	"	English.
300	$1\frac{1}{4}$	16	3342	186	1.9	355	312	Manilla	In use 2 years; no signs of wear.
250	2	12	1470	470	1.32	620	800	Cotton	
200	$1\frac{1}{4}$	6	3926	280	2.0	560	390†	R'whide	Worn out in 4 years.
200	1	12	3078	178	1.86	330	200	Manilla	

200	1	10	2400	275	1.8	495	250†	R'whide	In use 3 years; good condition yet.
150	1	13	2400	158	1.8	285	250†	"	
150	$1\frac{1}{4}$	6	2800	295	1.83	540	312	Manilla	
150	$\frac{7}{8}$	18	3190	82	1.86	152	190†	R'whide	In use $3\frac{1}{2}$ years; very little wear.
140	$\frac{3}{4}$	8	2480	230	1.8	415	112	Manilla	Three ropes worn out in 2 years.
100	$\frac{7}{8}$	5	4333	152	2.1	320	190†	R'whide	Worn out in 3 years.
80	$1\frac{1}{8}$	7	2513	150	1.8	270	252	Manilla	In use 3 years; good condition.
60	$\frac{3}{4}$	2	3600	275	1.9	540	112	"	Would last from 4 to 6 months.
60	2	5	1592	246	1.7	415	800	Hemp	German.
50	$1\frac{1}{8}$	1	1800	920	1.72	1,600	252	Manilla	Broke 3 times 5 months; replaced in 5 months.
50	$\frac{5}{8}$	4	2513	164	1.8	295	78	"	Would last 6 weeks to 4 months; finally taken out.
45	$1\frac{1}{4}$	5	1700	175	1.72	300	312	"	
40	2	1	5000	265	1.66	490	800	Cotton	In use 8 years; good condition yet.
35	$\frac{7}{8}$	9	1318	97	1.7	164	153	Manilla	In use $3\frac{1}{2}$ years; no signs of wear.
30	$\frac{5}{8}$	3	2256	146	1.77	257	78	"	
25	$\frac{3}{4}$	1	3140	262	1.86	490	112	"	Wears out in 1 year.
20	$\frac{1}{2}$	4	3140	55	1.86	120	50	"	Quarter-turn drive.

* Calculated from $\frac{T_1}{T_2} = \epsilon^{\phi a(1-z)}$, ϕ is assumed equal to 0.5 for cotton, and 0.3 for well-greased manilla and rawhide. The arc of contact on smaller pulley was taken as 165 degrees in all cases.

† The assumed allowable strain for rawhide ropes was taken at $250d^2$ pounds. No distinction was made between manilla and cotton.

In determining the driving force, $P\left(=\frac{33{,}000 \times \text{H.P.}}{V}\right)$, no allowance was made for engine resistance, which will reduce the above values about 10 per cent for the larger engines, but this refinement is unnecessary in view of the fact that engines are frequently run at a higher power than their normal rating.

available for transmission of power. To allow for possible imperfections in the rope, due to its manufacture or the material used, we must allow a further reduction of, say, one fifth (that is, 10 per cent of the original strength of an unlubricated, unspliced rope); thus we have only 40 per cent of the original breaking strength, which we may consider as the actual working strength. Allowing 10 per of this as the apparent working strain (equal 4 per cent of original strength), we obtain 36 per cent as the actual margin for wear and unknown stresses which may be set up in the rope; that is, we have practically only an actual factor of safety of ten, used in its ordinary acceptation, instead of twenty-five, as would appear from the low working stress.

The working strain in executed rope transmissions will be found to vary considerably, as shown in Table V; but plants which have been successful, as well as those in which the wear of the rope was destructive, indicate that $200d^2$ pounds is an economical working strain.*

As this value is such a small percentage of the breaking strength, it is unnecessary to use a coefficient varying with the diameter of rope, as the difference is not worth considering. From data furnished by the Messrs. Pearce Brothers, of Dundee, who have erected rope-belting extensively, Prof. Unwin shows that in different cases in practice the driving force, or difference in tension, on the two portions of the rope is equal to 75 to 80 d^2 pounds; that is,

$$T_1 - T_2 = P = 75 \text{ to } 80\ d^2,$$

where T_1 = tension on driving side;
T_2 = tension on slack side;
P = driving force.

---

* See paper by C. W. Hunt, Trans. A. S. M. E., vol. XII. page 230.

It is probable that the value of the ratio $\dfrac{T_1}{P}$ varies from 1½ to 2½ in ordinary practice, depending upon the speed of the rope, the coefficient of friction, and the arc of contact between rope and pulley.

Unwin assumes $\dfrac{T_1}{P} = 1.2$ when the belt embraces 0.4 of the circumference of the smaller pulley; hence the greatest tension would be

$$T_1 = 1.2P = 90d^2 \text{ to } 96d^2.$$

This is based upon a coefficient of friction $= 0.7$ for a $45°$ groove, which we believe to be greatly in excess of its average value for running ropes. Moreover, the rope velocity was not considered in determining the above ratio; at speeds over 2000 feet per minute the influence of centrifugal force cannot be neglected, as it produces a very considerable force in the rope, and for these reasons the value of the ratio $\dfrac{T_1}{P}$ will be greater in average practice than 1.2.

If we assume a speed of 4000 feet per minute, the value of this ratio for greasy ropes may be taken equal to 2, from which there is obtained $T_1 = 2 \times 75d^2$ to $80d^2 = 150d^2$ to $160d^2$—a value somewhat less than that which we have assumed as a suitable working strain, viz., $200d^2$ pounds. In a recent communication from Messrs. Combe, Barbour & Combe of Belfast, who have been engaged in furnishing rope transmission for thirty years, it is stated that their basis of calculating the horse-power is to assume that a rope 5½ inches circumference (1¾ inches diameter) working on a four-foot pulley going 100 revolutions per minute will drive 8 h. p. under medium circumstances; and by "medium circumstances" they mean "that the ropes must work at a distance of at least 20 feet from centre of shafts and at a less inclination than 40° from the horizontal, at a speed not under 2000 feet per minute. Should the ropes be working vertically or at an angle greater than 45°, instead of taking 8 h. p. as the basis, you should take 7 or 6 respectively, according as the conditions grow worse and worse. On the other hand, should the ropes work horizontally and at a greater distance than 20 feet from centre to centre, and with a speed up to 3600 feet per minute, and the pulleys be fairly large, say from 5 to 7 feet in diameter, you may take 10 instead of 8 as the basis." From these considerations the working strain may readily be determined. If, as before, we take $\frac{T_1}{P} = 2$, we find that under average conditions the allowable working strain will be $T_1 = 135 d^2$ pounds. Under more favorable conditions and a higher velocity, $\frac{T_1}{P}$ will be greater and T_1 will approach $200 d^2$.

Although we have assumed the normal working load not to exceed $200 d^2$ pounds, this must be considered as the economical load for the lasting qualities of the rope; in many cases, however, the first cost, the more convenient adaptation of smaller ropes, and the use and lesser cost of smaller pulleys outweigh the greater economy obtained by the larger ropes, and loads are carried far in excess of that given. By the use of a greater number of wraps the working load on each will be reduced; but this adds to the first cost of the plant, and many concerns prefer to put in a new rope every year or two rather than put in more or larger ropes every six or eight years.

The ropes most commonly found in use vary from ⅝ inch to 2 inches in diameter, although other sizes are frequently employed; in one case, cited by Mr. T. S. Miller,[*] a rope

---

[*] Trans. A. S. M. E. 1891.

no larger than $\frac{6}{18}$ inch diameter is used very satisfactorily to transmit 20 h. p.

The largest rope in use for this purpose, so far as the writer is aware, is $3\frac{1}{4}$ inches in diameter: it is of cotton (Lambeth), and is used to drive the cable drums in the Washington Street power station of the North Chicago Street Railway Company.

In England and Germany ropes smaller than 1 inch in diameter are seldom employed except for machine driving; in ordinary cases the sizes most frequently adopted vary from $1\frac{1}{2}$ to 2 inches in diameter.

In main drives where a number of ropes are used, the tendency, as judged from recent practice, seems to favor a diameter about $1\frac{5}{8}$ or $1\frac{3}{4}$ inches.

With a given velocity and working tension the weight of rope required for transmitting a given horse-power will be the same, irrespective of the diameter of rope; the smaller rope will require more parts, but the weight will be the same. As we have stated, many engineers prefer to use a greater number of ropes over wide-faced pulleys; but in order to reduce the expense incident to a large number of grooves in the pulleys a closer margin is allowed on the smaller ropes, and in consequence the normal working strain on each rope is usually increased far in excess of that which would be necessary if the same weight of rope were employed as for the larger diameter. As a result of this, the small ropes are rapidly worn out, and frequent renewals become necessary.

We are aware that small ropes are advocated and installed by many engineers, but we believe this to be wrong both in principle and practice. In rope-driving the first cost and erection of small driving-pulleys may influence a designer to use small ropes; but as ropes are sold by the pound, and as the weight necessary to transmit a given power should be the same, irrespective of the diameter, it is evident that the first cost of the rope itself will be the same, whether large or small ropes are employed. Moreover, if the pulley is proportioned to the size of rope in each case, the smaller rope will last only about one third as long as a rope twice its size, under similar conditions.

CHAPTER VII.

However desirable it may be to use a given diameter of rope, the conditions of the problem frequently prohibit the employment of such ropes, and the designer must determine whether he shall use a smaller diameter or a different material or method of transmission.

If we assume that there is a minimum diameter of pulley which may be safely used for any given diameter and speed of rope, it will be evident that the number of revolutions of the pulley imposes conditions which limit the choice of rope diameter.

Thus if the maximum speed of rope be taken at 5000 feet per minute for a permanent installation, in which the working load is $200d^2$ pounds, and the least diameter of pulley $D = d^{1.7} \left(\sqrt[3]{V} \right) + 12''$ (page 179), then the greatest number of revolutions which can be obtained under these conditions with a 1-inch manilla rope will be 550.

If the least diameter of pulley for cotton ropes be taken equal to $0.8D$, then the greatest speed will be approximately 700 revolutions per minute for the same diameter of rope. If a greater rotative speed be desired, it is evident that a smaller rope must be used.

In any case, the greatest number of revolutions which may be obtained without excessive wear for a given diameter of rope will be found in Table VI, which has been determined from the formula $N = \dfrac{V}{\pi D}$, in which

$N =$ revolutions per minute of smaller pulley;
$V =$ velocity of rope in feet per minute;
$D =$ least permissible diameter of pulley.

For machine-driving greater speeds are obtained by the use of smaller ropes than those given in the table; but it is not advisable in most cases to use a rope less than $\frac{3}{4}$ inch diameter for general transmissions.

TABLE VI.—GREATEST REVOLUTIONS PER MINUTE FOR GIVEN DIAMETER OF ROPE.

Diameter of Rope.	Maximum Revolutions per Minute of Smaller Pulley corresponding to a Linear Velocity of 5000 Feet per Minute.	
	Manilla.	Cotton.
$\frac{3}{4}$	710	890
1	550	670
$1\frac{1}{4}$	430	530
$1\frac{1}{2}$	350	440
$1\frac{3}{4}$	280	350
2	240	290

The wear of a rope is both internal and external. As we have previously noted, the internal wear is due to the bending of the fibres and their sliding upon one another, which produces a grinding action,—very much increased when the strands are not lubricated or when a hard twist is given to the rope, thus preventing by the greater compression of the fibres upon one another that freedom of action which is so essential to the life of the rope. It is evident that a similar compression of the fibres will occur when a rope under tension is wrapped around a pulley: the greater this tension the greater also will be the compression and distortion of the fibres.

The external wear is due to the contact between the rope and the sides of the groove in which it runs, and is greatly increased when slip occurs; roughness in the groove also increases the wear, and for this reason the rim should be turned smooth and polished, as the outer fibres, rubbing on a rough-turned or cast surface, will gradually break, fibre by fibre, and thus give the rope a short life.

Contact between different ropes, or between a rope and some obstructing surface, such as a partition wall, post, or floor-beam, is frequently the cause of a large portion of the external wear of a rope: this may be due to faulty construction or erection; pulleys designed with too small a pitch between the grooves, or running out of true, causing the ropes to vibrate and flap against each other, or, as in outside work, a swaying, producing contact, may be set up in the ropes, due to the action of the wind.

Excessive swaying will also tend to cause the rope to jump its groove. In order to prevent this and reduce the side motion as much as possible it is often customary in outdoor drives to place idlers for both tight and slack sides of the ropes so as to guide each portion as it enters upon or leaves the groove.

A characteristic uniform surging sometimes occurs in flying ropes due to the harmonic vibration which is set up when the speed and distance between shafts bears a certain relation to the tension in the ropes. Cases of excessive vibration due to this cause have been remedied by slightly increasing or decreasing the speed of the ropes. Where such vibration causes the rope to beat against an obstruction, as a floor or ceiling, the external wear is of course increased.

With the same total stress in a rope, it may be assumed that the wear, both internal and external, increases directly with the number of flexures, the slip, and the surface in contact; and also, that a reverse bending is more injurious to the rope than single bends in a constant direction. For a given speed the number of flexures and the actual surface in contact with the pulleys will decrease as the distance between centres increases, and hence the wear will vary inversely with the distance between centres of pulleys, but it must be noted that with imperfect construction an increased distance between shafts will favor swaying and rubbing of the ropes against each other and the edges of the grooves.

The number of flexures and the surface in contact will evidently increase directly as the velocity, and therefore the wear may be assumed to vary directly as the velocity of the rope. If we assume that two 1¼-inch ropes are necessary to transmit a given horse-power, it will require eight ⅝-inch ropes to transmit the same power at the same speed and tension per square inch of section. If suitable pulleys are used in each case the wear will be considerably greater with the smaller ropes. For the total external surface of the eight ropes in contact with the pulley each revolution is twice as great as that obtained with the 1¼-inch ropes; moreover, the distance between centres of shafts will generally be considerably less with the smaller ropes, and as the number of revolutions of the smaller pulleys should be more than twice as great for the same speed of rope (since the pulleys are less than half the size of those used for the larger rope; see page 180), the number of flexures and the wear of the rope on the pulleys will be greater with the ⅝-inch rope; for not only is each rope bent more than twice as many times per minute, thus producing eight times the bending in the smaller ropes, but, as the slip is independent of the diameter of the rope, it will be evident for the same proportional stress and arc of contact that the slip will be four times greater in the case of the smaller ropes, even if we neglect the differential driving effect which may be assumed to increase with the number of ropes.

Thus it will be seen that under similar conditions and proportional stress, we should expect the smaller rope to wear out more than twice as fast as one double its size; and when the stress is proportionally greater in the smaller rope, as we ordinarily find it, the wear will be still greater.

These conclusions are borne out in practice, for in transmissions using small ropes, ¾-inch in diameter and under, the life of manilla ropes is usually only from six to twelve months; in many cases such ropes will last only three months, although others have been in active service for periods varying from one to two years.

On the other hand, the larger-sized ropes, one inch to two inches in diameter, will last from two to six years, and under favorable conditions large ropes have lasted eight and even ten years.

The same is true regarding cotton ropes.

The comparatively short life of small ropes used in machine-driving has led to the belief that cotton ropes wear out rapidly; but such an impression, at least regarding the larger sizes, is altogether erroneous, as these ropes when properly put on and cared for will give good service for ten or twelve years.

Messrs. John Musgrave & Sons, Bolton, Eng., who have had a large experience with rope-driving, state that some of their ropes have been in use for seventeen years and were still in good order.

The life of a rope, whether of manilla or cotton, will depend altogether upon the work it has to do and the attention it receives.

Two ropes cut from the same coil can be put to work on different drives, and one will last only six months while the other will be in good condition after continual service for ten years.

From an investigation of numerous examples of rope-driving under a variety of existing conditions the writer is led to believe that ropes less than ¾ inch diameter should not be used if it is at all practicable to employ the larger sizes, and that ropes one inch in diameter and over are to be preferred where it is possible to use the larger pulleys which are necessary for such ropes.

With larger ropes the wear is not only much less, but, where the usual multiple-wrap system is used, when a number of yarns give way the rope does not part at once if subjected to a greater or sudden tension, but may run until a convenient opportunity offers to shut down; whereas with the small rope, having a greater number of wraps, when a strand or a number of yarns give way, any increased stress due to additional load or imperfections in the system is liable to still further rupture the yarns in the weaker rope, and a sudden break or pulling out occurs.

Besides the greater life of the rope, and consequent less cost, the saving of time on account of fewer breakdowns and stoppages is a factor worth considering; and although this feature does not cut as much of a figure with rope-driving as it does with factory belting, yet it is of such importance that many men would rather use some other form of transmission than suffer the annoyance incident to resplicing a broken rope every few months. While the time lost in repairing the rope or in laying down a new one may not be great in itself, the stoppage of a department in a busy season may prove to be a serious loss. Where foundries and isolated shops have been driven by small ropes this has happened so frequently that the ropes have been taken out and replaced with shafting, or other more expensive method of transmission.

In order to avoid any serious loss of time or inconvenience due to a sudden rupture two independent ropes should be used, each having its own tension sheave and weight. This does not involve any more wraps than would otherwise be used for a single wind, for the normal working stress is in any case so much less than the actual strength that for a temporary run of a few hours, or even days, one rope could readily carry double its working load in case the other should give out. When two independent ropes are thus used they may be wound separately, each wrap occupying successive grooves on the pulleys; or, which is more frequently the case, the ropes may be wound in parallel, thus bringing each rope in alternate grooves: in either case it is preferable to use an independent tightener, although a single-tension carriage, provided with two sheaves, may be used; but with this latter, owing to local causes and the difficulty of splicing both ropes of an equal length, the load is not as well distributed between the two ropes.

In subsequent considerations of the driving power of ropes the relation between the ultimate strength, weight per foot of length, normal working strain, and the diameter of rope will be represented by the following equations which have been determined for manilla transmission rope:

Let d = diameter of rope in inches;
w = weight of rope in pounds per foot;
S_1 = breaking strain in pounds;
t = normal working strain in pounds;
x = an empirical coefficient.
Then $w = 0.3216 d^2$;
$S_1 = 100 d^2 x$;
$t^* = 200 d^2$;
$x = 81 - 9d$.

The weight w per foot of length varies considerably in different makes of rope, depending upon the amount of twist and the foreign matters in the rope.

It is well known that much of the cheaper manilla rope

---

* On account of the relatively great difference between S_1 and t, it is not thought advisable to consider the increase in strength of the smaller ropes, as in any case the difference would be very slight, and, moreover, it must be noted that t is the normal estimated strain, and may vary considerably from the actual strain.

on the market is largely adulterated with weighing material, such as gelatine size, French clay, and white lead. Thus some manilla ropes will weigh not more than $0.26d^2$ pounds when dry and very loosely twisted; in other cases the weight will be as much as $0.46d^2$ pounds per foot of length. The value we have given, viz., $0.32d^2$ pounds, corresponds very closely to the average weight of good quality lubricated transmission ropes.

Cotton ropes are about twenty per cent lighter for equal diameters, and will vary from $0.20d^2$ to $0.29d^2$ pound per foot. In the following table (VII) $0.32d^2$ has been used for manilla and $0.26d^2$ for cotton ropes.

TABLE VII.—WEIGHT OF ROPES.

Diameter of Rope.	Weight in Pounds per Foot.	
	Manilla. $w = 0.32d^2$.	Cotton. $w = 0.26d^2$.
¾	0.18	0.15
1	.32	.26
1¼	.50	.40
1½	.72	.58
1¾	.98	.79
2	1.28	1.04

CHAPTER VIII.

In determining the horse-power which a rope will transmit under given conditions the centrifugal force due to the velocity and weight of rope is an important factor, and its influence should be considered in all cases where the speed is greater than 2000 feet per minute; for at high velocities this force diminishes the pressure exerted between the rope and the circumference of the pulley, thus reducing the friction between rope and pulley. When, therefore, a rope has to transmit a given force, P, it must be subjected to a greater tension the greater the centrifugal force F_o. At a speed of about 90 feet per second the centrifugal force increases faster than the power from increased velocity of the rope, and at 140 feet per second this force equals the assumed allowable back tension in the rope; and since the transmitting force is equal to the difference in tension in the two parts of the rope, it will be seen that no power will be transmitted at this speed unless the assumed allowable tension be exceeded.

It is evident that for a given total tension the less back tension required to prevent the slip of the rope on the pulley the greater will be the power transmitted at a given speed.

The determination of this back tension is, however, attended with a degree of uncertainty, as there are no conclusive experiments which give reliable data for its calculation; the coefficient of friction, ϕ, as stated by various authorities, varying all the way from 0.075 up to 0.88.*

---

* The probable reason for such widely divergent values lies in the fact that the coefficient of friction varies with the percentage of slip, and those tests made with very little slip would show a small

Reuleaux quotes the experiments of Leloutre and others as indicating a value of 0.075 for cylindrical pulleys with new hemp rope, 0.088 for semicircular grooves, and 0.15 for a wedge groove of 60°.

Experiments by the Messrs. Pearce Brothers, of Dundee, give a value of ϕ equal to 0.57 to 0.88 for ropes on ungreased pulleys, and $\phi = 0.38$ to 0.41 when the pulleys are greased.

Unwin states that the coefficient of friction for ropes on a flat-metal pulley is equal to 0.28, from which the actual coefficient for a grooved pulley is obtained by multiplying 0.28 by the cosec. of half the angle of the groove. For an angle of 45° this would give $\phi = 0.72$. These latter values are probably very much higher than is ordinarily found in actual practice with well-lubricated ropes and moderate slip. From a consideration of the above and various other experiments, and the conditions under which they were carried out, it would appear that for ropes which are partly worn and sufficiently greased to wear well with a low percentage of slip a value of 0.12 for a flat-surfaced, smooth-metal pulley will approach very closely to those conditions which obtain in average practice, from which the following working coefficients are deduced:

$$\phi = 0.12 \operatorname{cosec}\left(\frac{\text{Angle of groove}}{2}\right)$$

Angle of groove..........	30°	35°	40°	45°	50°	55°	60°
Coefficient of friction, ϕ...	0.46	.40	.35	.31	.28	.26	.24

Besides varying with the angle of groove, as shown, the coefficient; on the other hand, it is safe to assume that the larger values were obtained under conditions in which the slip was greatly increased. Varying atmospheric conditions and different degrees of lubrication are also largely responsible for these divergent results.

See paper by Prof. Lanza on "Friction of Leather Belting," in Trans. A. S. M.E., vol. VII. p. 347; also "Experiments on Power Transmitted by Belting," by Wilfred Lewis, in Trans. A. S. M. E. vol. VII. p. 549.

coefficient of friction is affected by the condition of the rope, and for dry ropes ϕ may be taken somewhat greater than the above value; ϕ will also be increased with an increased percentage of slip.

If the arc of contact on smaller pulley, the coefficient of friction between rope and sheave, and the total tension in the rope be known, the tension on slack side of the pulley, and hence the horse-power transmitted, can be readily determined from the following considerations:*

Assuming, as before, that the driving force P is equal to the difference in tension T_1 on the driving side of the rope and T_2 on the driven side, and noting that the driving force must equal the friction F between the surfaces, we obtain

$$T_1 - T_2 = P = F.$$

The friction F depends upon the arc of contact α between the rope and its sheave, the coefficient of friction ϕ, and upon the centrifugal force F_0 set up in the rope, due to its velocity and weight; it is, however, independent of the diameter of pulley. To determine the values of F, T_1, and T_2 it will be necessary to assume a given tension in the rope; also its speed and weight, coefficient of friction, and arc of contact.

Let $f =$ friction between element of rope and pulley;
$g =$ acceleration due to gravity $= 32.16$;
$p =$ normal pressure exerted by element on pulley;
$t =$ allowable working tension $= 200d^2$ pounds;
$v =$ velocity of rope in feet per second;
$w =$ weight of rope per unit length and area $= 0.3216d^2$ pounds per foot;
$z =$ abbreviation for $\dfrac{w}{g}\dfrac{v^2}{t}$;

---

* Weisbach, vol. III. p. 254. See also Reuleaux.

A = area of cross-section of rope;
F_e = centrifugal force due to speed and weight of rope;
P = driving force = $T_1 - T_2$;
R = radius of pulley;
T = tension in rope at any point;
T_1 = tension in rope on tight side;
T_2 = tension in rope on slack side;
α = least arc of contact between rope and pulley—circular measure = 0.0175 × arc in degrees;
ϕ = coefficient of friction.

If in Fig. 42 T is the tension in the rope at any point D, then the tension at the point E, whose distance from D is ds, will be $T + dT$. Assuming f to represent the

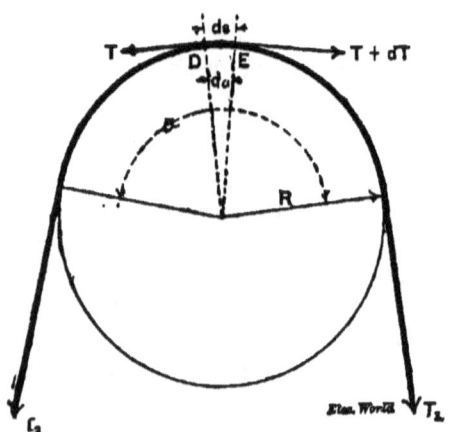

FIG. 42.

friction on the element DE, we shall have $f = dT$ when the several forces are in equilibrium. Since the friction is dependent upon the normal pressure p exerted by the element upon the rim of the pulley, and since this normal pressure is diminished by the centrifugal force due to the weight and velocity of the element, we shall have

ROPE-DRIVING. 115

$$dT = \phi(p - F_0). \quad \ldots \ldots \quad (4)$$

Now p is the resultant of the two forces T and $T + dT$; hence

$$p = T \sin \frac{d\alpha}{2} + (T + dT) \sin \frac{d\alpha}{2},$$

which may be assumed equal to $Td\alpha$ on account of the smallness of $d\alpha$ and dT.

The centrifugal force of the element of the rope is

$$F_0 = wA \frac{v^2}{Rg} ds = w \frac{T}{t} \frac{v^2}{Rg} ds;$$

and since $ds = Rd\alpha$, we have

$$F_0 = w \frac{T}{t} \frac{v^2}{g} d\alpha.$$

Substituting these values of p and F_0 in $dT = \phi(p - F_0)$, we obtain

$$dT = \phi \left(Td\alpha - w\frac{v^2}{g} \frac{T}{t} d\alpha \right). \quad \ldots \quad (5)$$

For convenience, let $z = \dfrac{wv^2}{gt}$; then

$$dT = \phi T(1 - z) d\alpha, \quad \text{or} \quad \frac{dT}{T} = \phi(1 - z) d\alpha.$$

Integrating,

$$\int_{T_2}^{T_1} \frac{dT}{T} = \int_0^\alpha \phi(1 - z) d\alpha,$$

$$\text{hyp log } \frac{T_1}{T_2} = \phi(1 - z)\alpha; \quad \ldots \quad (6)$$

hence

$$\frac{T_1}{T_2} = \epsilon^{\phi\alpha(1 - z)} \quad \text{or} \quad T_1 = T_2 \left[\epsilon^{\phi\alpha(1 - z)} \right],$$

in which ϵ is the base of the hyp log $= 2.7183$; therefore

$$T_1 - T_2 = P = T_2\left[\epsilon^{\phi a(1-z)}\right] - T_2 =$$
$$T_2\left[\epsilon^{\phi a(1-z)} - 1\right]. \quad \ldots \quad (7)$$

If we assume $\dfrac{T_1}{T_2} = r$, there is obtained

$$\frac{r}{r-1} = \frac{T_1}{T_1 - T_2} = \frac{T_1}{P}.$$

These ratios, r and $\dfrac{r}{r-1}$, Reuleaux calls the friction modulus and the stress modulus, respectively.

As the common logarithm of ϵ is 0.43430, the value of r may be more readily obtained from common log $r = 0.4343\,\phi a(1-z)$; if the arc of contact is given in degrees, $a = 0.0175\alpha°$, which gives the common log of

$$r = 0.4343 \times .0175\alpha°\phi(1-z) = 0.007578\phi\alpha\,(1-z). \quad (8)$$

As the weight of a manilla rope one foot long $= 0.32d^2$ pounds, the value of z for varying speeds can be determined from

$$z = \frac{wv^2}{gt} = \frac{0.01d^2v^2}{t}. \quad \ldots \quad (9)$$

If now we assume a constant working stress $t = 200d^2$ pounds, then

$$z = \frac{0.01d^2v^2}{200d^2} = 0.00005v^2. \quad \ldots \quad (10)$$

In the work which follows we shall assume that the tension T_1 in the rope on the tight side (driving tension) equals the allowable tension t.

From these assumptions the following table (VIII) of the values of $1 - z$ has been computed:

ROPE-DRIVING. 117

TABLE VIII.—VALUES OF $1-z$ FOR A WORKING STRESS EQUIVALENT TO $200d^2$ POUNDS.

Velocity in Feet per Minute.	Values of $1-z$.	Velocity in Feet per Minute.	Values of $1-z$.
1000	0.98	5500	0.58
2000	0.94	6000	0.50
2500	0.91	6500	0.41
3000	0.87	7000	0.32
3500	0.83	7500	0.22
4000	0.78	8000	0.11
4500	0.72	8500	0.0
5000	0.65		

It will be seen from the above that when the velocity of the rope is as great as 8500 feet per minute, $1-z=0$, hence $\log r = 0$ and $\dfrac{T_1}{T_2} = 1$; that is, $T_1 - T_2 = 0$, and therefore no power will be transmitted unless the assumed working tension t be exceeded.

In average work the lesser arc of contact embraced by the rope—generally on the smaller pulley—will be about 165°, and this value may be assumed for approximate calculations with a working degree of accuracy. If the angle in degrees is known, its value, α, in circular measure, can be obtained from Table IX, in which $\alpha = 0.0175 \alpha°$.

TABLE IX.—ANGLE EMBRACED BY ROPE.

Degrees, $\alpha°$.	Circular Measure, α.	Fraction of Circumference, $360/\alpha°$.	$\dfrac{\alpha°}{360°}$
105	1.83	0.29	
120	2.09	0.33	
135	2.35	0.37	
150	2.62	0.42	
165	2.88	0.46	
180	3.14	0.50	
195	3.43	0.54	
210	3.66	0.58	
240	4.19	0.66	

If we now assume the coefficient of friction ϕ to be 0.31 for a 45° groove, we may obtain the value of the expressions

$$\frac{T_1}{T_2} = e^{\phi a(1-z)} = r \quad \text{and} \quad \frac{T_1}{P} = \frac{r}{r-1}.$$

In order to simplify calculation, the following table (X) contains values of r and $\dfrac{r}{r-1}$, which will enable the horse-power transmitted by a rope to be determined with a degree of accuracy depending upon the assumption of the coefficient of friction:

TABLE X.—FRICTION AND STRESS MODULI.

$\phi a(1-z)$.	$r = \dfrac{T_1}{T_2}$.	$\dfrac{r}{r-1} = \dfrac{T_1}{P}$.
0.1	1.11	10.41
0.2	1.23	5.40
0.3	1.35	3.86
0.4	1.49	3.02
0.5	1.65	2.54
0.6	1.82	2.22
0.7	2.01	1.99
0.8	2.22	1.82
0.9	2.46	1.69
1.0	2.72	1.58
1.1	3.00	1.50
1.2	3.32	1.43
1.3	3.67	1.37
1.4	4.06	1.33
1.5	4.48	1.29

The following application will show the use of the tables. Let it be required to determine the horse-power transmitted by a rope 1 inch in diameter running at a velocity of 4000 feet per minute over a pulley with 45° grooves. Assuming an arc of contact of 165°, we find from Table IX $\alpha = 2.88$; for the required velocity, 4000 feet per minute, Table VIII gives 0.78 as the value of $1 - z$; therefore, assuming the coefficient of friction $\phi = .31$, we obtain

$$\phi a(1-z) = 0.31 \times 2.88 \times .78 = .69.$$

From Table X the value of $\dfrac{T_1}{P}$, corresponding to 0.69, is about 1.99; and as $T_1 = 200d^2$ pounds $= 200$ in the present case, we find $P = \dfrac{T_1}{1.99} = 100$ pounds. Since $\dfrac{PV}{33000} =$ h. p., there is obtained h. p. $= \dfrac{100 \times 4000}{33000} = 12.15$, represented by the ordinate lm in Fig. 43.

The loss due to centrifugal force may now be obtained

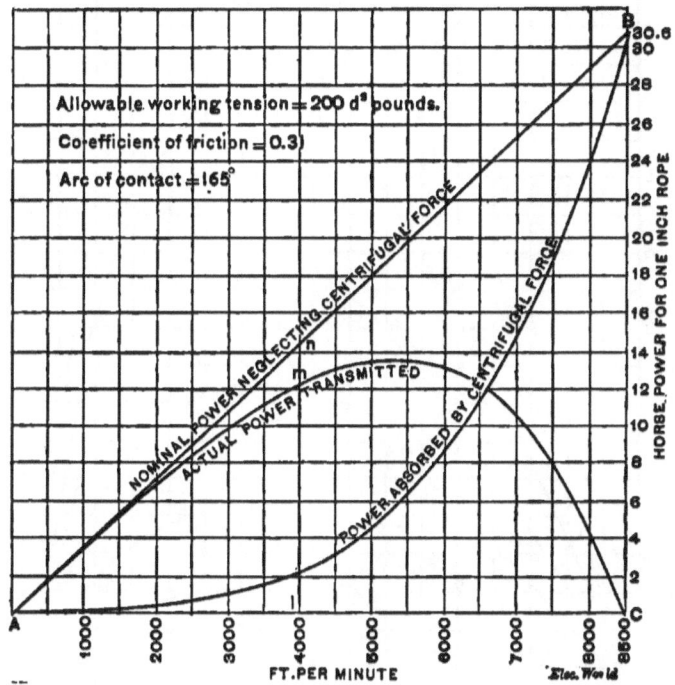

FIG. 43.—CENTRIFUGAL EFFECT IN ROPES.

by assuming the latter reduced to zero, in which case the factor $1 - z$ is equal to unity; therefore $\log r = .434\phi\alpha$, from which, with the previous conditions, we obtain $P = 120$ and the corresponding h. p. $= 14.5$.

This value is represented on the diagram, Fig. 43, by the ordinate ln: the difference between ln and $lm - mn$ will then be the loss due to the centrifugal force set up in the rope = 14.5 − 12.15 = 2.35 horse-power.

For any special case where the data are known or may

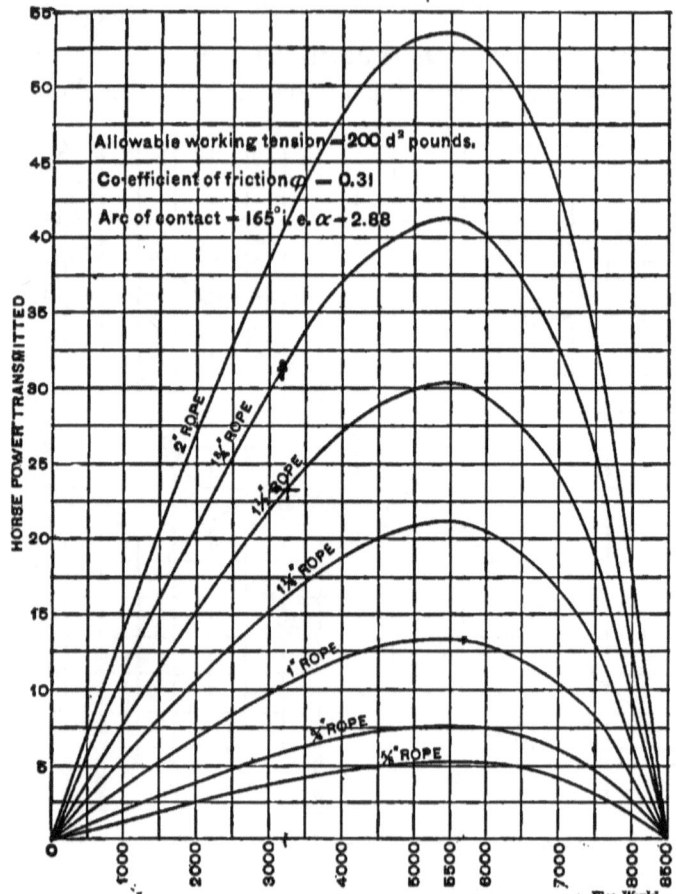

Fig. 44.—Horse-power transmitted by Manilla Ropes.

be determined, the formulas and tables already given should be used to ascertain the horse-power transmitted, or the diameter and number of ropes required for a certain work, as the case may be. For average work, however, it will be

ROPE-DRIVING. 121

found that the assumed values of α and ϕ, previously noted, will give very satisfactory results, and upon these assumptions the writer has computed the following table of horse-power (Table XI) for various-sized ropes, running at speeds from 1000 to 7500 feet per minute:

TABLE XI.—HORSE-POWER TRANSMITTED BY ROPES.

Working Strain $= 200d^2$ pounds.

$d =$ diameter of rope in inches.

Velocity of Rope in Feet per Minute.	Diameter of Rope.						
	⅝	¾	1	1¼	1½	1¾	2
1000	1.24	2.25	3.57	5.59	8.02	10.85	14.20
2000	2.70	3.84	6.84	10.68	15.39	20.93	27.36
2500	3.30	4.71	8.38	13.10	18.86	25.66	33.54
3000	3.83	5.46	9.80	15.39	21.87	29.74	38.88
3500	4.30	6.23	11.09	17.33	24.94	34.03	44.35
4000	4.74	6.83	12.15	18.98	27.33	37.17	48.59
4500	5.01	7.24	12.89	20.15	29.00	39.45	51.57
5000	5.20	7.47	13.29	20.76	29.89	40.65	53.15
5500	5.29	7.60	13.53	21.14	30.43	41.39	54.11
6000	5.08	7.32	13.10	20.36	29.32	39.77	52.12
6500	4.74	6.83	12.13	19.00	27.34	37.21	48.63
7000	4.12	5.93	10.54	16.47	23.72	32.26	42.18
7500	3.25	4.67	8.32	13.00	18.73	25.42	33.23

The graphic representation of these values, Fig. 44, shows the effect of centrifugal force in diminishing the power transmitted under an assumed working tension, and would indicate that with tensions of $200d^2$ pounds the speed should not exceed 5500 feet per minute. The increasing effect and loss of power due to centrifugal force in the rope can also be seen in the diagram, Fig. 43, which represents the horse-power transmitted by an inch rope under an assumed constant tension of 200 pounds. The straight line AB shows the power which would be transmitted if centrifugal force were neglected, and is obtained by making $z = 0$ in the general equation

$$\log r = 0.4343\alpha\phi(1-z);$$

the curve AC represents the power transmitted when centrifugal force is taken into account; and the curve AB shows the power absorbed by centrifugal force: this latter curve is obtained by subtracting the vertical ordinates between the straight line AB and the curve AC. By a considation of the diagram it will be seen that at speeds less than 2000 feet per minute the power absorbed by centrifugal force is very small, and may be neglected as far as practical results are concerned. Beyond this speed, however, the loss from this cause increases very rapidly, until, as previously shown, at a speed of about 8500 feet per minute the whole of the allowable tension is absorbed.

Assuming that the maximum power is transmitted by a rope at a velocity of about 5500 feet per minute, it is evident that the first cost of the rope will be a minimum for a given power when running at this speed. The ratio of the first cost of the rope running at any other speed may be obtained by dividing the horse-power at 5500 per minute by the horse-power at the required speed.*

Thus, if the first cost of a 1¼-inch rope which will transmit 21.14 h. p. at 5500 feet per minute be represented by unity, the cost at 3000 feet per minute will be

$$\frac{21.14}{15.39} = 1.38,$$

since a 1¼-inch rope running at 3000 feet per minute will transmit 15.39 h. p.

The relative first cost for a given diameter of rope to transmit the same horse-power at varying speeds is shown in the accompanying Table XII.

Although the first cost of a rope to transmit a given horse-power is a minimum for a speed of about 5500 feet per

---

* C. W. Hunt, in Trans. A. S. M. E., vol. XII.

minute, yet the economy is not as great as would appear from the foregoing table, for the effect of wear must be considered. The causes of wear, internal and external, have been previously discussed; it will be sufficient to note here that the

TABLE XII.—RELATIVE FIRST COST OF ROPE-DRIVING.

Velocity of Rope in Feet per Minute.	Relative First Cost per Horse-power.	Velocity of Rope in Feet per Minute.	Relative First Cost per Horse-power.
1,000	3.78	4,500	1.05
2,000	1.89	5,000	1.02
2,500	1.62	5,500	1.00
3,000	1.38	6,000	1.03
3,500	1.22	6,500	1.12
4,000	1.10	7,000	1.28

internal destructive effect produced by bending and distorting the fibres and the wear due to external contact, slipping, or wedging in the grooves of the pulleys, may, within the limits of ordinary practice, be considered as directly proportional to the velocity of the rope. What this wear is in terms of the velocity there is not sufficient data to determine, but if the coefficient be represented by c, the *relative* wear for a given diameter may be determined by multiplying the velocity by this coefficient; that is, the relative wear $= cv$, in which the wear increases directly with the velocity, but not, however, directly with the horse-power transmitted.

If we assume the coefficient to be such that the wear on a rope at 1000 feet per minute is equal to unity, then the wear on the rope at any other speed will be

$$\text{Relative wear} = \frac{\text{required speed}}{1000}.$$

To determine the relative wear per horse-power transmitted by a given rope at varying speeds, it will be necessary to determine the wear per horse-power for the rope running at any required speed, and then divide this value by the wear per horse-power when running at 1000 feet per minute. Let it be required to determine the wear of a rope transmitting a given horse-power at 5500 feet per minute, as compared to what it would be at 1000 feet per minute.

The horse-power transmitted at 1000 feet per minute is found to be 3.57 for a one-inch rope, at which speed we have assumed that the wear is equal to unity, hence the wear per horse-power may be considered equal to

$$\frac{1}{3.57} = .28.$$

At a speed of 5500 feet per minute the horse-power transmitted by the same rope is 13.53, but the wear at this speed is assumed to be five and a half times greater than at 1000 feet, other conditions being the same, therefore the wear per horse-power $= \frac{5.5}{13.53} = .406$; that is, the wear of any rope transmitting one horse-power at 5500 feet per minute is $\frac{.406}{.28} = 1.45$ times the wear which would occur at 1000 feet per minute.

From this it will be seen that although the first cost of a rope is cheaper at the higher speeds, the rope lasts longer while running at the lower speeds—conditions remaining constant.

Taking the case we have been considering, it is found that the relative first cost of the rope is inversely as the horse-power transmitted, or

$$\frac{\text{Cost of rope at 1000 feet}}{\text{Cost of rope at 5500 feet}} = \frac{13.53}{3.57} = 3.78;$$

that is, the first cost is 3.78 times greater for the slower speed. But it is shown above that the rope will wear out nearly 50 per cent faster at the increased speed; therefore, taking the life of the rope as well as the first cost into consideration, the relative cost to transmit a given horse-power at the speeds noted will be $\frac{3.78}{1.45} = 2.6$ times greater for the lower speed. This can be determined more conveniently by assuming $H =$ horse-power transmitted at V feet per minute, $H_1 =$ horse-power transmitted at V_1 feet per minute, where $H_1 > H$ and $V_1 > V$. The wear per horse-power in each case will then be proportional to $\frac{V}{H}$ and $\frac{V_1}{H_1}$; the relative first cost of the rope per horse-power running at the lesser speed, compared to that when running at the greater, will be as H_1 is to H; hence the ratio of relative cost, taking wear into account, is

$$\frac{H_1}{H} \times \frac{\frac{V}{H}}{\frac{V_1}{H_1}} = \frac{V}{V_1}\left(\frac{H_1}{H}\right)^2 \quad \ldots \quad (11)$$

TABLE XIII.—RELATIVE WEAR AND COST OF ROPE PER HORSE-POWER TRANSMITTED.

Velocity in Feet per Minute.	Relative Wear per Horse-power transmitted.	Relative Cost of Rope per H. P. transmitted, considering Wear.
1,000	1.	1.
2,000	1.03	.54
2,500	1.06	.45
3,000	1.10	.40
3,500	1.13	.36
4,000	1.18	.347
4,500	1.25	.345
5,000	1.34	.36
5,500	1.45	.38
6,000	1.64	.44
6,500	1.93	.52
7,000	2.40	.80
7,500	3.22	1.36

Table XIII has been calculated upon the above basis of comparison, namely, that the wear in a rope at 1000 feet per minute is equal to unity and increases directly as the speed; and also, that the cost of ropes for a permanent

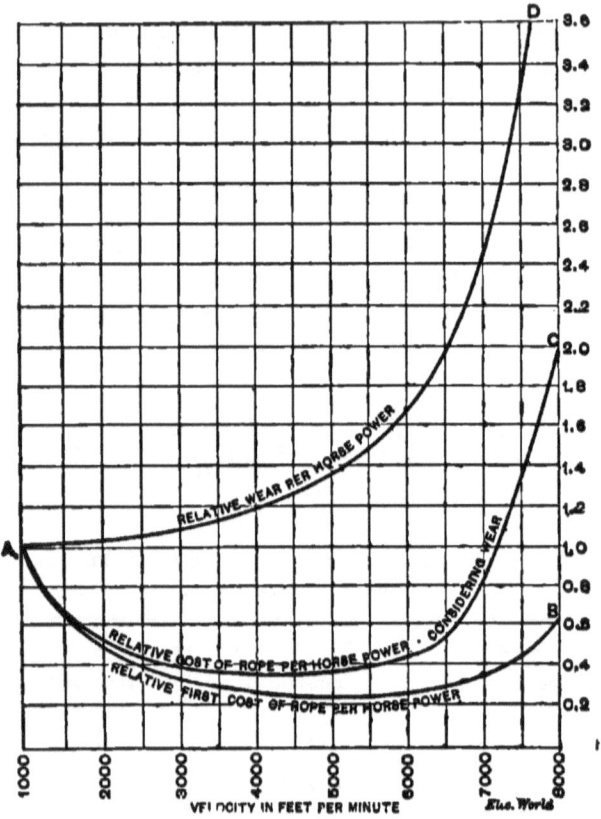

FIG. 45.—RELATIVE WEAR AND COST OF ROPE PER H. P.

installation is proportional to the square of the ratio of the power transmitted at different speeds multiplied by the inverse ratio of the corresponding speeds. To ascertain the relative wear per horse-power of a rope running at any given speeds, it will only be necessary to form a ratio between the values in the table corresponding to the given speeds. Thus the wear per horse-power at 5000 feet per minute, as compared to that at 2500 feet, will be $\frac{1.34}{1.06} = 1.26$.

In the same way the relative ultimate cost per horse-power of a rope running at these speeds will be $\frac{.36}{.45} = 0.8$, or 20 per cent less for the greater speed. The accompanying diagram, Fig. 45, represents these relative values graphically, to which is added the curve of relative first cost. It will be noticed that although the first cost of a rope is a minimum for a speed of 5500 feet per minute, when wear is considered the minimum cost occurs at a speed of 4500 feet per minute.

CHAPTER IX.

As previously noted, it is desirable in all cases of rope transmission to so arrange the drive that the slack side of the rope shall be on the upper part of the pulley, thus increasing the arc of contact, as the two sides will then approach each other when in motion.

In order that the desired tensions T_1 and T_2 shall be attained in the two parts of a rope, the deflections or sag must be of predetermined values. The centre line of the rope will lie in a curve, which may be determined with no appreciable error by assuming the rope to have no elasticity and to be of constant cross-section, under which condition the curve will be that known as the catenary, the transcendental equation of which is

$$y = \frac{c}{2}\left(\epsilon^{\frac{x}{c}} + \epsilon^{-\frac{x}{c}}\right). \quad \ldots \quad (12)$$

Let the form of curve in which the rope hangs be represented by $PO'P'$ (Fig. 46), in which O' is the lowest point of the rope. Take O' as the origin of coördinates, and let $x = O'A'$ and $y' = A'P$ be the abscissa and ordinate of any point P in the curve, and let l be the distance between the points of support.

Since we have assumed the rope to be perfectly inelastic, the tension at any point of the curve must be in the direction of the rope. Let T be the stress in the rope at P, the vertical and horizontal components of which are represented by V and H respectively. Assume the length of curve $O'P = s$. Since the weight of a unit length of rope $= w$, the vertical component of the tension T is evidently equal to one half the weight of the rope between the points P and P', or $V = \tfrac{1}{2}w \; \overline{PO'P'} = ws$. To determine the length of curve, produce the tangent through the point P and rectify the curve $O'P$ on this tangent, making $s = PQ$.

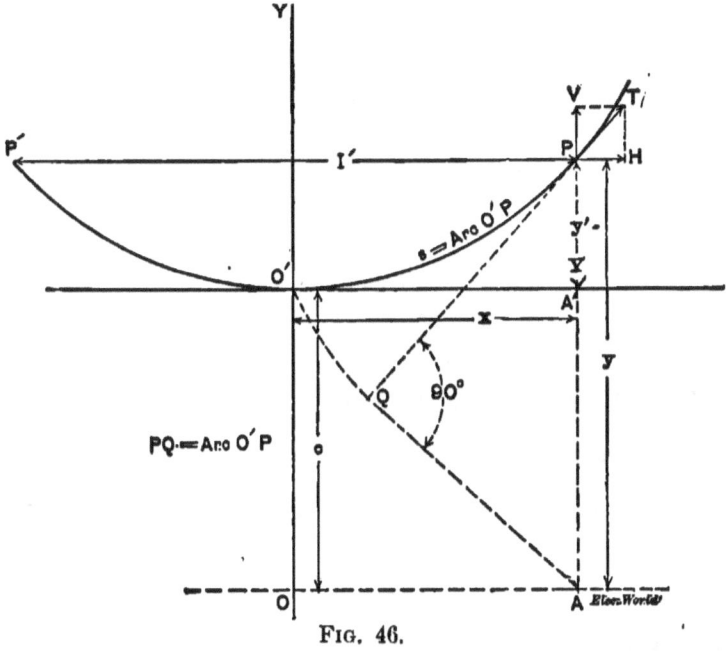

Fig. 46.

Erect a perpendicular to PQ at the point Q, meeting the vertical from P at A; then will OA drawn through A parallel to $O'A'$ be the directrix to the catenary, and QA will equal $AA' = c$.* Hence

$$\overline{PQ}^2 = \overline{PA}^2 - \overline{QA}^2.$$
$$s^2 = (c + y')^2 - c^2\,; \quad \therefore \; s = \sqrt{(y')^2 + 2cy'}.$$

Therefore $\quad V = ws = w\sqrt{(y')^2 + 2cy'}.$

---

* For geometry of this curve see "The Funicular Polygon" in Bowser's "Analytic Mechanics," p. 216 *et seq.*; also Price's "Mechanics," vol. I.

At the vertex O the tension is horizontal and equal to the weight of a length, c, of the rope; but this tension is the same as the horizontal component at the point P; hence $H = wc$. Since $T = \sqrt{V^2 + H^2}$, we obtain the following value for the tension in the rope:

$$T = w\sqrt{y'^2 + 2cy' + c^2} = w(c + y'). \quad . \quad . \quad (13)$$

In order to determine the parameter, c, let the equation of the curve $y = \dfrac{c}{2}\left(\epsilon^{\frac{x}{c}} + \epsilon^{-\frac{x}{c}}\right)$ be developed into the following series:

$$y = \frac{c}{2}\Bigg(1 + \frac{x}{c} + \frac{x^2}{1.2\,c^2} + \frac{x^3}{1.2.3\,c^3} \cdots + 1 - \frac{x}{c}$$
$$+ \frac{x^2}{1.2\,c^2} - \frac{x^3}{1.2.3\,c^3} + \cdots \Bigg).$$

Since the character of the curve is such that the quotient $\dfrac{x}{c}$ is a proper fraction, the series will be converging. Stopping at the third member as giving sufficient accuracy, we have

$$y = \frac{c}{2}\left(2 + \frac{x^2}{c^2}\right) = c + \frac{x^2}{2c},$$

therefore $\qquad x^2 = 2c(y - c). \quad . \quad . \quad . \quad . \quad . \quad (14)$

Substituting the value of $y = y' + c$ in this equation, we obtain $x^2 = 2cy'$, which is the equation of a parabola referred to its axis and the tangent to its vertex. Now let $\dfrac{l}{2}$ (Fig. 47) equal the half distance between supports $= x$, and $h =$ sag of the rope $= y'$; then from the previous equation we obtain $\left(\dfrac{l}{2}\right)^2 = 2ch$:

hence the parameter $c = \dfrac{l^2}{8h}$. Substituting this value of

c in equation (13), we have

$$T = w \left(\frac{l^2}{8h} + h \right),$$

in which w equals the weight of a unit length of rope $= 0.32d^2$ for manilla; that is, the tension at any point in the rope is equal to the weight of a portion equivalent in length to the parameter plus the ordinate y' of the point. From this equation we may obtain the sag of the driving or driven portions of the rope by substituting for T the values of T_1 and T_2, the deflections corresponding to which may be represented by h_1 and h_2 (Fig. 47). T_1 will evidently be con-

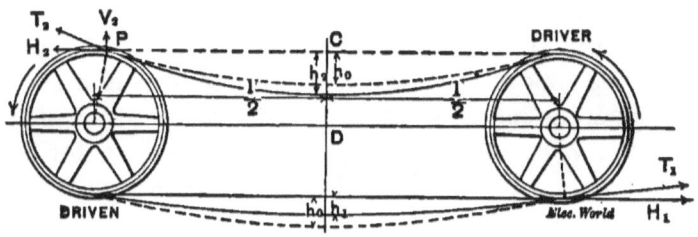

Fig. 47.

stant and equal to $200d^2$ if our preconceived conditions are maintained, but the value of T_2 will be variable, increasing with the speed.

Practically, it will be impossible to maintain a constant tension in the rope, so that the amount of sag obtained by calculation is liable to vary with the conditions of service. The tension may, however, be approximately determined by the deflection. Assuming the distance between supporting points of the rope equal to the distance between centres of pulleys, and solving for h, we obtain

$$h_1 = \frac{1}{2} \frac{T_1}{w} \pm \frac{1}{2} \sqrt{\frac{T_1^2}{w^2} - \frac{l^2}{2}};$$

$$h_2 = \frac{1}{2} \frac{T_2}{w} \pm \frac{1}{2} \sqrt{\frac{T_2^2}{w^2} - \frac{l^2}{2}}.$$

The positive and negative signs before the radical in these equations indicate two values for h, the lesser of which only is to be used. As pointed out by Reuleaux, between the two lies a value $h = \frac{1}{2}\frac{T}{w}$, which is obtained when the quantity under the radical $= 0$; that is, when $T = \frac{wl}{\sqrt{2}}$. This deflection is interesting, as it denotes the minimum stress which may exist in the rope.

Since the sum of the tensions increases with the speed, the sag of the rope when at rest is not directly obtainable from the previous values of h_1 and h_2.

The tension necessary for adhesion, which constitutes a part of the stress, T_1, is dependent upon the speed at which the rope is intended to be run, so that in order to determine the sag h_0 when at rest, for a given maximum tension T_1, the initial tension must be obtained for any given speed, and this value substituted in the general formula

$$h = \frac{T}{2w} \pm \frac{1}{2}\sqrt{\frac{T^2}{w^2} - \frac{l^2}{2}}.$$

Assuming the tension on the tight side of the rope to be made up of three parts, namely, the driving force P, the centrifugal force F_0, and the tension T_3, necessary to balance the strain for adhesion, we obtain

$$T_1 = P + F_0 + T_3.$$

In like manner the tension on the slack side of the rope may be assumed to be produced by the strain necessary for adhesion, plus the strain due to centrifugal force; that is,

$$T'_2 = T'_3 + F_0.$$

It is evident that if the normal tension T'_1 be diminished by the centrifugal force the remaining stress will be equal to the tension necessary for adhesion plus the driving force; hence if we obtain the value of the ratio $\dfrac{T_1}{T_2}$ from $\epsilon^{\phi a}$, in which the stress due to centrifugal force is neglected, we shall have in T_2 the initial tension necessary for adhesion, or $T'_2 = T_2$. If T'_1 is assumed to be constant, then T'_2 will be constant for a given coefficient of friction and arc of contact; therefore $\dfrac{T'_1 + T'_2}{2}$ will equal the tension in the rope when the latter is at rest.

The following simpler method for obtaining the sag, though less exact, is sufficiently accurate for any practical case that may arise, for it must be borne in mind that any theoretical calculation for the deflection of a running rope can at best be only an approximation, as it is exact only when the rope is running at its normal speed, transmitting its full load and strained to its normal tension. Let l be the distance between two shafts which are at the same

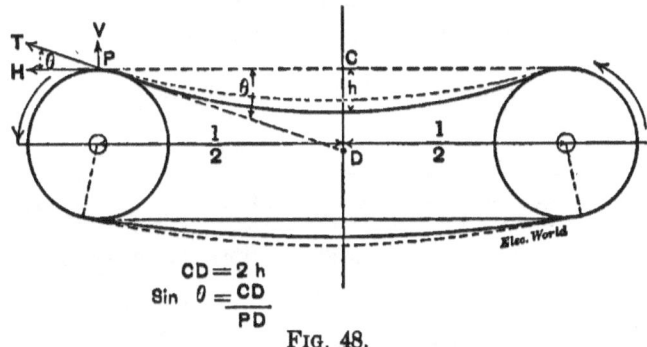

FIG. 48.

level (Fig. 48), and let h be the deflection of the rope; also, let $CPD = \theta$ be the inclination of the rope at P, from which

$$\sin \theta = \dfrac{CD}{PD} = \dfrac{2h}{\sqrt{\left(\dfrac{l}{2}\right)^2 + (2h)^2}} = \dfrac{4h}{\sqrt{l^2 + (4h)^2}};$$

and hence, since V is the vertical component of the tension

T in the rope and equal to $\dfrac{W}{2}$, where W is the weight of rope between supports, we shall have

$$T = \frac{V}{\sin \theta} = \frac{W}{8h}\sqrt{l^2 + (4h)^2} = W\sqrt{\frac{l^2}{64h^2} + \frac{1}{4}h^2}.$$

As h is small compared to l, we may neglect $\dfrac{1}{4}h^2$, as it will have no appreciable influence on the result; then if we assume the length of rope to be equal to the distance between the shafts, we shall have, approximately,

$$T = \frac{Wl}{8h} = \frac{wl^2}{8h},$$

where $wl = W$; hence the sag of the rope at the centre will be

$$h = \frac{wl^2}{8T}, \quad \cdots \cdots \quad [15]$$

in which $w = 0.32d^2$ for manilla ropes.

From this formula the following table (XIV) has been

TABLE XIV.—DEFLECTION OF ROPE.

Calculated from $h = \dfrac{wl^2}{8T}$.

Distance between Pulleys in Feet.	Deflection in Feet in Slack Side of Rope when $T_2 = T_1 - P =$				Deflection in Tight Side of Rope when $T_1 = 200d^2$ Pounds.	Deflection in Both Sides of the Rope when Initial Tension $T_0 = 141d^2$.
	$86d^2$	$91d^2$	$100d^2$	$112d^2$		
	Corresponding to Velocity in Feet per Minute of:					
	2000	3000	4000	5000		
30	0.42	0.39	0.36	0.32	0.18	0.25
40	0.74	0.70	0.64	0.57	0.32	0.45
60	1.67	1.58	1.44	1.28	0.72	1.02
80	2.97	2.81	2.56	2.28	1.28	1.82
100	4.65	4.40	4.00	3.57	2.00	2.84
120	6.70	6.33	5.76	5.14	2.88	4.10
140	9.12	8.61	7.84	7.00	3.82	5.58
160	11.90	11.25	10.24	9.14	5.12	7.27

ROPE-DRIVING.   135

computed and the curves plotted, as shown in Fig. 49, T_1 being assumed as constant and equal to $200d^2$ pounds. The tension T_2, which varies with the speed, has been separately determined for the several cases considered from

$$\frac{T_1}{T_2} = e^{\phi a(1-z)},$$

in which the coefficient of friction, ϕ, $= 0.31$ and $\alpha = 2.88$; hence

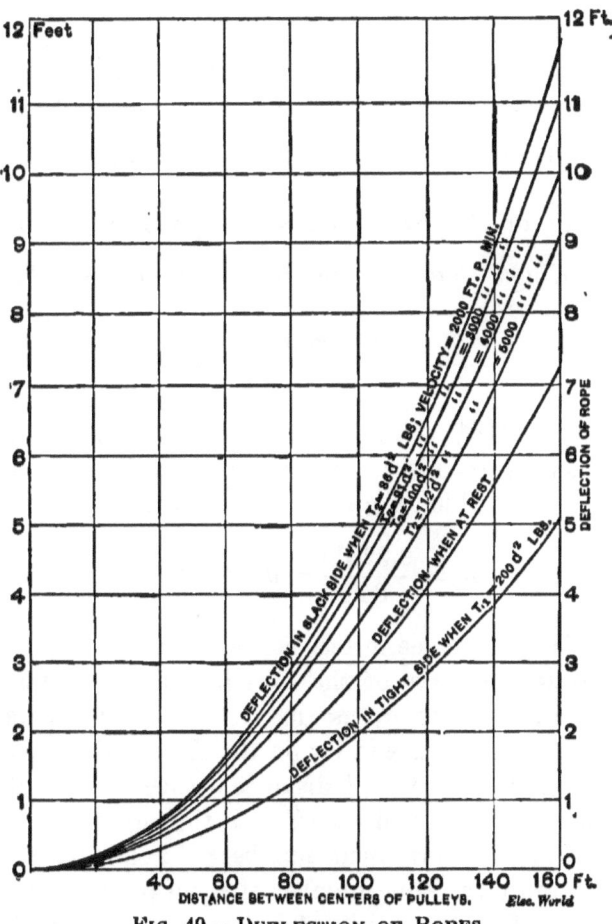

Fig. 49.—Deflection of Ropes.

$$\log \frac{T_1}{T_2} = 0.4343 \times 0.9(1 - z).$$

The deflection on the tight side will then be

$$h_1 = \frac{0.32 d^2 l^2}{8 T_1} = \frac{0.32 d^2 l^2}{8 \times 200 d^2} = 0.0002 l^2;$$

and on the slack side $h_2 = \dfrac{0.04 l^2}{T_2}$, in which T_2 has the values given in the table.

As previously pointed out, the initial tension in the rope when at rest may be obtained by neglecting F_0, in which case we have $\log \dfrac{T_1}{T_2} = 0.4343 \phi \alpha$, from which we find $\dfrac{T_1}{T_2} = 2.46$, or $T_2 = 82 d^2$; and since $T_2 = T_1$ when there is no centrifugal force acting, the initial tension T_0 will equal

$$\frac{T_1 + T_2}{2} = \frac{200 d^2 + 82 d^2}{2};$$

hence $T_0 = 141 d^2$. From this we may obtain the deflection when the rope is at rest, as noted in the last column of Table XIV, which has been computed from

$$h_0 = \frac{0.32 d^2 l^2}{8 T_0} = 0.000284 l^2,$$

where l is the distance in feet between centres.

To draw the curve of the rope in order to determine the space it will occupy, assume it to hang in a parabola with origin at vertex and lay off the X and Y axes, as in Fig. 50. Let $O5 = \tfrac{1}{2} l = \tfrac{1}{2}$ distance between points of suspension of the rope, and let $5H = h$, the sag of rope. Divide $5H$ into a convenient number of equal parts, also divide $O5$ into the same number of equal parts; erect perpendiculars from the points of division 1, 2, 3 ... and join

1', 2', 3'... with the origin O. The curve drawn through the points of intersection will represent one branch of the parabola desired.

In outdoor drives, where the configuration of the ground would prevent the proper amount of sag being used, or, in

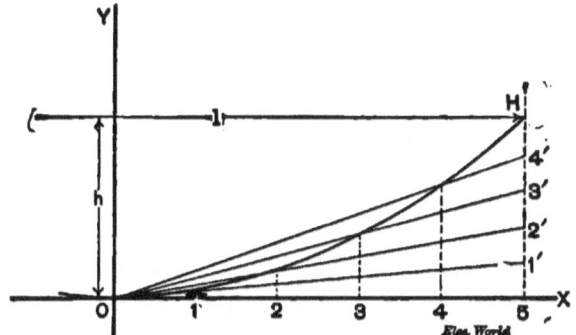

FIG. 50.—METHOD OF LAYING OUT CURVE.

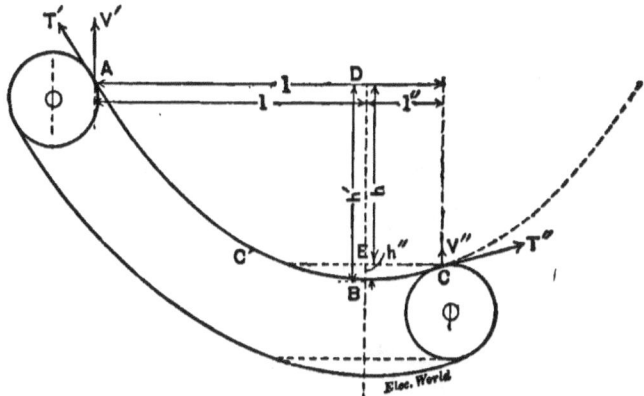

FIG. 51.—INCLINED TRANSMISSION.

general, where obstructions prohibit the proper deflection for a given distance between centres of shafts, the rope should be carried on supporting pulleys or sag-carriers, examples of which have already been given.

Where the pulleys are placed at different heights we have an inclined transmission, and the curve in such cases is unsymmetrical, as in Fig. 51.*

This curve is best solved by an approximation similar to that already given.

It will be noticed that the curve of the rope ABC is made up of two unequal parts, AB and BC, whose horizontal projections are l' and l'' and their corresponding deflections h' and h''. The given difference in height between the points of support of the rope is $ED = h = h' - h''$, and their known horizontal distance is $l = l' + l''$. From the preceding formula for the deflection (equation [15]) we have

$$h' = \frac{wl'^2}{8T'} = \beta l'^2 \quad \text{and} \quad h'' = \frac{wl''^2}{8T''} = \beta l''^2,$$

in which β is taken as a coefficient depending upon the value of the tension in the rope $= \dfrac{0.32}{8T}$.

Since $h = h' - h''$ and $l = l' + l''$, we have

$$h = \beta(l'^2 - l''^2) = \beta(l' - l'')(l' + l'') = \beta l(l' - l'');$$

therefore $l' - l'' = \dfrac{h}{\beta l}$. But $l' = l - l''$ and $l'' = l - l'$; hence $2\beta l l' - \beta l^2 = h$, and the required distance $l' = \dfrac{h + \beta l^2}{2\beta l}$. In the same way $l'' = \dfrac{\beta l^2 - h}{2\beta l}$, and the deflections are

$$h' = \beta\left(\frac{h + \beta l^2}{2\beta l}\right)^2 \quad \text{and} \quad h'' = \beta\left(\frac{\beta l^2 - h}{2\beta l}\right)^2,$$

in each of which equations β has a separate value depending upon the tension in the rope. It is evident that the tension T' at A will be greater than that at C, on account of the greater weight of rope lying in the upper branch of

---

* Weisbach, vol. III. p. 243.

the parabola. If T' is known, the lesser tension T'' can be determined by assuming the length of the rope to be the same as if the points A and C, Fig. 52, were at the same height and l feet apart. From previous considera-

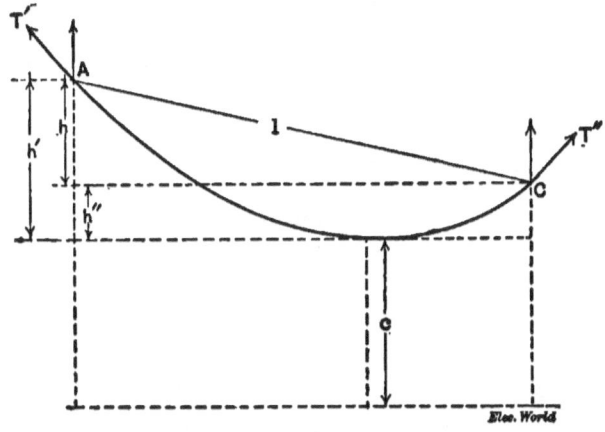

FIG. 52.

tion it is seen that the tension at any part of the rope in a parabola is equal to the weight of rope equivalent in length to the parameter c added to the ordinate of the point; hence if h'' equals the ordinate of $C = h' - h$, we shall have

$$T'' = w(c + h'') \quad \text{and} \quad T' = w(c + h').$$

Since $c = \dfrac{l^2}{8h'}$,

$$T'' = w\left(\dfrac{l^2}{8h'} + h''\right) \quad \text{and} \quad T' = w\left(\dfrac{l^2}{8h'} + h'\right);$$

therefore $T'' - wh'' = T' - wh'$. Substituting for h' its value, $h + h''$, there is obtained

$$T'' = T' - wh.$$

When a tension-carriage is used the necessary weight can be ascertained from the formula for back tension,

$T'_2 = T_2 + F_0$; if we assume $F_0 = 0$, we shall have the initial tension necessary for adhesion equal to the tension in slack side of rope, that is, $T_2 = T_2$. But $T_2 = \dfrac{T_1}{2.46}$ (from log $\dfrac{T_1}{T_2} = 0.434\phi\alpha$, assuming previous values of ϕ and α); therefore the initial tension in the rope will be $T'_2 = \dfrac{T_1}{2.46} = \dfrac{200 d^2}{2.46}$.

When diameter of rope equals

⅝ inch	the initial tension	=	31	pounds.
¾ "	" " "	"	=	45	"
⅞ "	" " "	"	=	62	"
1 "	" " "	"	=	82	"
1¼ "	" " "	"	=	127	"
1½ "	" " "	"	=	184	"
1¾ "	" " "	"	=	250	"
2 "	" " "	"	=	325	"

The actual weight to be placed on the tension-pulley will depend upon the arrangement of drive, and in order to obtain the best results should be determined for each particular case. It is to be noted that with a vertical tension-carriage the weight of carriage and sheave must be considered as a portion of the weight producing tension in the rope.

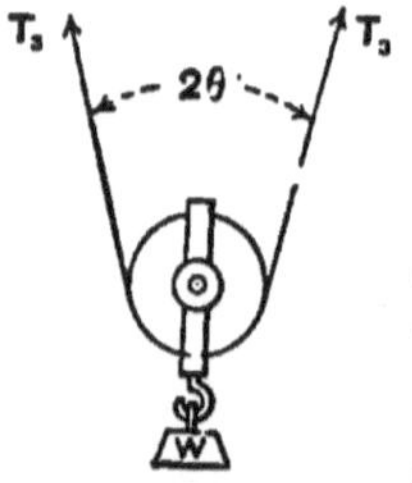

FIG. 53.

If W equals the total weight necessary to maintain the tension T_2 in each part of the rope leading off from the tension-pulley, then, when these two portions are parallel, we shall have $W = 2T_2$; but if the rope leads from the tension-pulley so that it includes an angle 2θ between its sides, as shown in Fig. 53, then the weight W will be less, and may be found from $W = 2T_2 \cos \theta$.

CHAPTER X.

It has been previously stated (see page 7) that where rope-driving is used the loss at the engine in friction may be taken in a general way as about 10 per cent of the rated horse-power of the engine, that an additional 10 per cent is absorbed by the shafting, and that from 5 to 8 per cent may be attributed to losses in the rope itself, due to resistance to bending, wedging in the grooves, differential driving effect, and creep, all of which affect the loss to a greater or less extent.

According to the accepted laws of solid friction we should expect that with an increased load on the engine the friction would be increased in direct proportion to the load, but in nearly all experiments to determine the friction of steam-engines we have the anomaly that the work necessary to overcome friction is practically constant, and, in fact, in many cases it is a little greater when running without load than when the engine is fully loaded.

However, it is probable that the ordinary laws of friction obtain here as in other cases of sliding and rolling contact; but instead of having a constant coefficient of friction for the surfaces in contact, the coefficient may be considered as a variable depending upon the degree and distribution of lubrication. The lubrication of engine bearings (both rolling and sliding contact), approaches more nearly to the condition which obtains when the bearings are subjected to an oil bath, and although the lubrication is restricted, yet the result is similar in action.

Experiments* on the friction of a well-lubricated journal

---

*Proc. I. M. E. November, 1885; also Kent's Mechanical Engineer's Pocket Book.

(oil bath), show that the absolute friction, that is, the absolute tangential force per square inch of bearing, required to resist the tendency of the brass to go round with the journal, is nearly a constant under all loads within ordinary working limits. Most certainly it does not increase in direct proportion to the load, as it should do according to the ordinary theory of solid friction. The results of these experiments seem to show that the friction of a perfectly lubricated journal follows the laws of liquid friction much more closely than those of solid friction. They show that under these circumstances the friction is nearly independent of the pressure per square inch, and that it increases with the velocity, though at a rate not nearly so rapid as the square of the velocity.

The experiments on friction at different temperatures indicate a great diminution in the friction as the temperature rises. Thus in the case of lard-oil, taking a speed of 450 revolutions per minute, the coefficient of friction at a temperature of 120° is only one-third of what it was at a temperature of 60°. In regard to engine friction, whatever be the cause, it is a well-known fact that the coefficient of friction decreases as the load increases, so that at all ordinary speeds the internal resistance of the engine may be considered sensibly constant, in which case the so-called friction-card of the engine represents practically the friction of the machine when fully loaded—the indicated power without load being sensibly the measure of the wasted work of the engine when in operation under load of whatever amount.

That is, the engine friction is independent of the load and is a function of the characteristic of the engine itself, of the speed of piston and rotation, and, to a slight extent, of the steam-pressure and of the method of steam-distribution: so that while we may speak of the friction as being a certain percentage of the horse-power of an engine, it must be understood to refer to the rated indicated horsepower;* at less than rated power the percentage of loss due to friction will be greater, and at maximum power the percentage will be less. This is shown by way of illustration in the following table (XV), which is taken from a paper presented by Prof. Thurston before the American Society of Mechanical Engineers in 1886:†

This engine, a "Straight Line," was 8 inches in diameter of cylinder by 14 inches stroke; it had a balanced valve with stroke varying from 2 to 4 inches according to position of governor and eccentric, a fly-wheel 50 inches in diameter, weighing 2300 pounds. Its rated load was 35 to 40 horse-power.

TABLE XV.—FRICTION PER CENT UNDER VARYING LOADS.

Revolutions per Minute.	Steam Pressure.	Indicator Horse-power.	Friction Horse-power.	Friction per cent.
232	50	7.41	3.35	45
229	65	7.58	2.60	34
230	63	10.00	4.00	40
230	73	11.75	3.65	32
230	75	14.02	4.02	28
230	80	15.17	3.17	21
230	75	16.86	2.86	17
230	75	28.31	3.36	11.75
229	60	33.04	3.16	9.5
229	58	37.20	2.34	6.3
229	70	43.04	3.19	7.4
230	85	47.79	2.75	5.8
230	90	52.60	2.60	4.9
230	85	57.54	2.54	4.4

Examining the above table, it will be seen that the friction of the engine varies somewhat with varying steam-pressures and total power, but in such a manner as to indicate the controlling cause, as, for instance, imperfect lubri-

---

* Thurston, Trans. A. S. M. E., vol. x. p. 110.
† Trans., vol. VIII. p. 90.

cation, to be irregular in action, and, possibly, to some extent, due to errors of observation and to accident. The average friction horse-power is 3.11 h. p., and the variations from this value are distributed throughout the whole series, showing that the work necessary to overcome the friction is practically constant, and independent of the load. The friction of this engine is quite low, as at its normal rating the percentage of loss is less than 7 per cent. According to D. K. Clark,* the frictional resistance of steam-engines varies from 8 to 20 per cent of their normal indicator h. p. —the size of engines experimented upon ranging from 13 to 350 h. p.

In recent tests of American engines it has been shown that with first-class workmanship and balanced valves the percentage of loss at normal working load may be reduced to about 6 per cent, both with Corliss and with high-speed single-cylinder automatic engines; but this is exceptional, and we may expect with ordinary lubrication that the frictional resistance will vary from 8 to 13 per cent of the normal load.

Compound engines of the better class should not absorb more than about 10 per cent, and triple-expansion engines no more than $12\frac{1}{2}$ per cent, under full load.

For small engines, either single or compound, there is probably little difference between the internal resistance, whether geared with ropes or flat leather belts, for the weights of fly-wheel and grooved pulley and the diameter of shaft would be essentially the same in each case. With larger engines, however, the belts would require a wider-faced fly-wheel, which, on account of the greater distance between bearings, would necessitate a larger shaft, and hence increased work in overcoming journal-friction—assuming the same weight of wheels and speed of rotation.

---

* "The Steam Engine," vol. II. p. 616.

ROPE-DRIVING. 145

With the rope-driving, also, the speed of the rim will be greater, and a somewhat lighter pulley may be used to insure the same degree of steadiness in running; in many cases the ropes are delivered from the fly-wheel in a nearly vertical direction, so that a certain portion of the weight on the bearings is neutralized by the upward pull of the ropes. Moreover, the elasticity and recoil of the ropes act

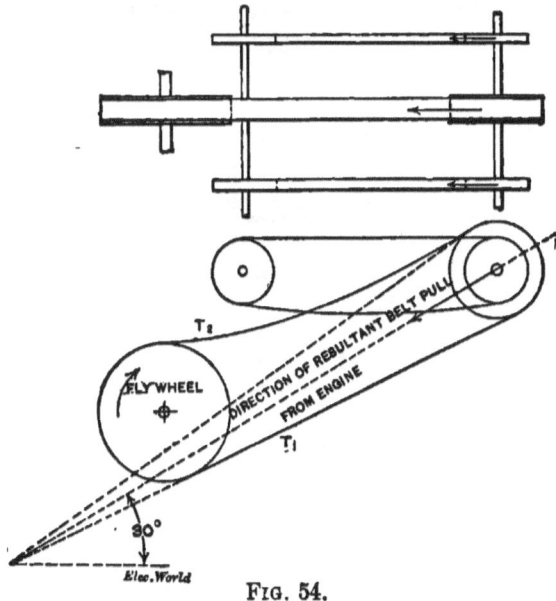

FIG. 54.

in the same manner as mass in the rim, and for this reason a lighter wheel may be used with rope-driving provided the construction is such as will permit it. The journal-friction is therefore presumably less in the larger class of engines employing ropes, when compared with those using belts. The difference is, however, not great; and since the actual resistance in any case is also dependent upon peculiarities of type and construction, the lower values of engine-friction previously given, viz., 8 to 10 per cent of normal horse-power, may be accepted as held good for rope-driving plants of medium size, while 9 to 12 per cent of the normal horse-power may be taken as suitable for the larger class of engines running under favorable conditions.

In the ordinary transmission of power by shafting we find the shaft loaded with pulleys and the power taken off in varying amounts throughout its entire length; it is unusual except in short lengths to receive the power at one end and transmit it at the other. Moreover, in long shaft-

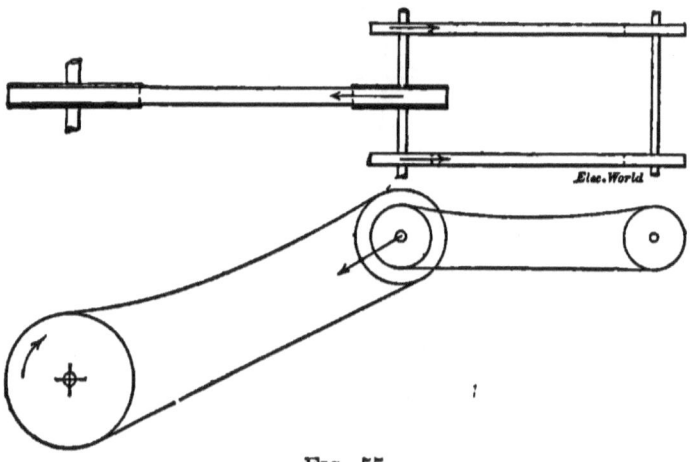

Fig. 55

ing the head or receiving shaft is usually situated midway between the ends, and the power distributed more or less uniformly from this head shaft to either end; therefore, in estimating the power absorbed by friction in ordinary mill or factory shafting loaded with pulleys the previous formulæ (page 70) do not apply, as these relate only to those cases where power is taken off at the end of the shaft.

The conditions of practice as we find them in actual transmissions are so various, that it is difficult to lay down any general rule by which the power absorbed by friction may be determined: the number and weight of pulleys and couplings, the intensity and direction of belt-pull, the condition of bearings and their lubrication, all affect the amount of work lost in friction.

For the ordinary factory shafting, from which power is taken fairly uniformly throughout its length and distributed horizontally to counter- or auxiliary shafts situated on one or both sides of the main shaft, there will be three general cases to be considered, as shown in Figs. 54, 55, and 56, and each of these cases will be modified, depending upon the direction of the belt to and from the main shaft.

For our present purposes it will be sufficient to take that case in which the shaft friction is a maximum for the assumed direction of main belt-pull corresponding to the arrangement shown in Fig. 54.

The friction will evidently be proportional to the weight of the shaft and the unbalanced belt-pull acting on the shaft.

The weight of pulleys, belts, clutches, and couplings carried by the line-shaft will vary from about one and one-half to three times the weight of shaft, so that the total weight on the bearings will vary from two and one-half to four times the weight of shaft; for head and jack shafts the total weight will probably vary from three to five times the weight of shaft.

In addition to this weight there is the unbalanced belt-pull, which increases the load on the bearings. Although the tension on the tight side of the belt should not ordinarily exceed about twice the tension in the slack side necessary for adhesion, yet it is probable that belts are frequently run with a ratio of tension equal to one to three, and occasionally one to four; on the other hand, it is a very common thing for belts, especially short ones, to be laced so taut that the initial tension is greatly in excess of that required for adhesion, in which case the sum of the tensions approaches twice that in the tight side of the belt.*

With ordinary shop-worn belting it will be safe to assume that the tension T_2 on the slack side of the belts is one half the tension T_1 on the tight or driving side, that is, $T_2 = \dfrac{T_1}{2}$; hence, since $T_1 - T_2 = P$, the driving force, we have

$$\frac{T_1}{2} = P, \quad \text{and} \quad HP = \frac{T_1}{2} \times \frac{V}{33000}.$$

The velocity of intermediate belting is so variable that

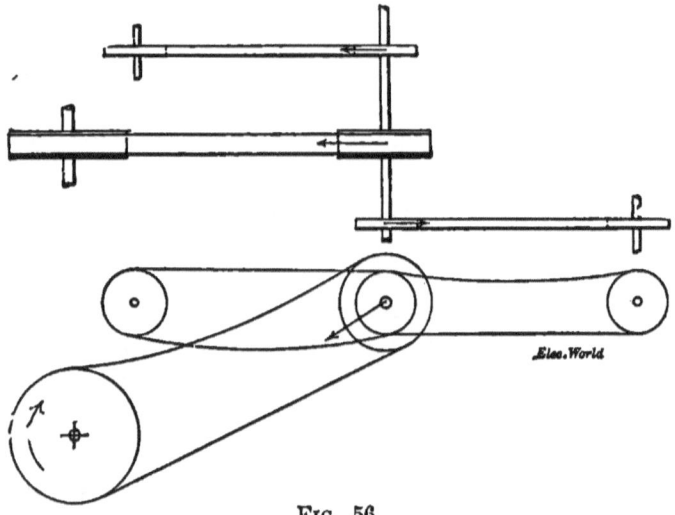

Fig. 56.

any assumption of speed must be regarded as applying to a particular case or representative of a certain type of factory, and cannot be taken as general. In many ma-

---

* This is often a source of much trouble, as the increased tension not only increases the loss due to friction but in many instances the useful power transmitted is not sufficient to drive the machine. In such cases, by slacking out the lacing or inserting a short piece of belt so as to reduce the tension, heavy cuts can readily be taken when it is practically impossible to run the machine empty with the tight belt.

chine-shops the average speed of intermediate belts is not more than 500 feet per minute; in others the average speed is more than twice as great, and in wood-working shops it is still greater.

For our present purpose we shall assume an average speed of 660 feet per minute for belts running from the main shaft to a secondary or counter shaft.

Substituting this value in $HP = \frac{T_1}{2} \times \frac{V}{33000}$, there is obtained $T_1 = \frac{2 \times 33000}{660} HP = 100 \times HP$; but since the horse-power which the shaft is capable of transmitting may be considered equal to $\frac{d^3 N}{100}$, where d is the diameter of shaft in inches and N the number of revolutions per minute, we have the tension on the tight side of all the belts $\Sigma T_1 = 100 \times \frac{d^3 N}{100} = d^3 N$; therefore the sum of tensions $\Sigma(T_1 + T_2) = \frac{3}{2} d^3 N$, and the pull per foot of length of shaft $= \frac{3 d^3 N}{2 L}$.

In the present case it will be noted by reference to Fig. 54 that there is an additional pull on the bearings due to the tensions in the belt from fly-wheel to main-line shaft. If the ratio of tight to slack side tension remains the same as before, and we consider that the velocity of main belt is four times as great as that of the intermediate belting, the additional belt-pull will equal approximately one fourth of the sum of the belt-pulls from the main to the counter shafts or machinery. The resultant of these tensions, combined with the weight of shafting and pulleys, will be the effective load on the bearings.

Assuming an angle of 30° with the horizontal for the line of action of the resultant pull on the bearings due to the tensions in the tight and slack sides of the main belt, the combined horizontal pull on the bearings will be

$$\cos 30° \times \frac{1}{4}\left(\frac{3}{2}d^2N\right) + \frac{3}{2}d^2N = 1.2\left(\frac{3}{2}d^2N\right);$$

and the vertical pull will be, when W_s = weight of loaded shaft,

$$W_s + \sin 30° \times \frac{1}{4}\left(\frac{3}{2}d^2N\right).$$

Therefore the resultant of both horizontal and vertical forces acting on the bearings will be

$$W = \sqrt{\left[W_s + \sin 30° \times \frac{1}{4}\left(\frac{3}{2}d^2N\right)\right]^2 + \left[1.2\left(\frac{3}{2}d^2N\right)\right]^2}.$$

As previously shown, the horse-power necessary to overcome journal-friction will be $HP_0 = \frac{Fv}{33000}$, where F is the force of friction at the circumference of shaft and v is the speed in feet per minute of a point on the circumference. If the bearing is well worn and fitted to its shaft, the resistance due to friction will probably lie between the limits $\frac{\pi}{2}\phi W$ and $\frac{4}{\pi}\phi W$, where ϕ is a coefficient which, from the results of experiments on shafting with ordinary lubrication, we have assumed equal to 0.06, and W is the resultant load, in pounds, on the bearings.

From the lesser of these values there is obtained

$$F = \frac{4}{\pi}\phi W. \quad . \quad . \quad . \quad . \quad . \quad (16)$$

But we have assumed that the weight of a loaded shaft varies from two and one-half to four times the weight of shaft; taking an average value of three for line-shafting, and noting that the weight of shaft per foot of length equals $\frac{\pi}{4}(3.36d^2)$, we have the friction on a loaded shaft L feet long, due to its weight =

$$\frac{4}{\pi}\phi 3\left(\frac{\pi}{4} \times 3.36d^2\right)L.$$

Substituting the value of W in formula (16) when the belt tensions are taken into account, and noting that $W_s = 3\frac{\pi}{4}(3.36d^2L)$, we have for the total friction load

$$F = \frac{4}{\pi}\phi\sqrt{\left[3\frac{\pi}{4} \times 3.36d^2L + \sin 30° \times \frac{1}{4}\left(\frac{3}{2}d^3N\right)\right]^2 + \left[1.2\left(\frac{3}{2}d^3N\right)\right]^2}$$

$$= \sqrt{[0.6d^2L + 0.014d^3N]^2 + [0.14d^3N]^2}.$$

From the formula for the power absorbed by friction we obtain

$$Fv = HP_o = \frac{F\pi dN}{33000 \times 12} \quad \text{or} \quad HP_o = 0.0^5 8dNF;$$

hence the ratio of power absorbed by friction to the horse-power which the shaft is capable of safely transmitting will be

$$\frac{HP_o}{HP} = \frac{0.0^5 8dNF}{0.01d^3N} = \frac{0.08F}{d^2} \text{ per cent.} \quad . \quad . \quad (17)$$

From this expression the following table (XVI) has been computed for the given diameters and lengths of shafting running at 100 and 250 revolutions per minute, the belt speed for the secondary belts being assumed at an average of 660 feet per minute.

For intermediate belts having a greater average velocity than that assumed, viz., 660 feet per minute, the friction horse-power for a given number of revolutions will be less than that given in the table. Thus if the average velocity of cross-belts equals 2640 feet per minute, the horse-power transmitted being the same, it follows that the tensions in the secondary belts will be one-fourth of that obtained with the lesser speed; if the main driving-belt have the same velocity, the tension in this belt may be considered equal to that existing in the intermediate belts: therefore, as the velocity of the belt increases for a given speed of rotation the sum of the tensions acting on the bearings will decrease, and the maximum horse-power transmissible by the shaft will be exerted with a decreasing friction loss.

TABLE XVI.—POWER ABSORBED BY FRICTION IN LINE-SHAFT.

Diameter of Shaft in Inches.	Revolutions per Minute.	Percentage of Loss when Length in Feet =			
		100	200	300	400
2	100	5.5	10.1	....	....
	250	7.8	11.6	....	....
2½	100	5.8	10.3	15	....
	250	8.9	12.4	16.6	....
3	100	6.1	10.4	15	19.6
	250	9.9	13.	17.2	21.5
3½	100	6.8	10.6	15.4	20
	250	11.4	14.2	18	22

Line of action of resultant of main-belt tensions = 30°.
Velocity of main belt 2640 feet per minute.
Belts from line shaft are horizontal and run at an average of 660 feet per minute.
All the belts are assumed to drive from one side of the shaft toward the engine.
Weight on bearings three times weight of bare shaft.

If in any case the shafts are belted vertically or at any other angle than that assumed, the formula for F will be modified accordingly.

For a head or jack shaft carrying heavier pulleys the weight acting on the bearings may be taken equal to four times the bare weight of shaft; in which case, other conditions remaining the same, we obtain

$$F = \frac{4}{\pi}\phi\sqrt{\left[\frac{4\pi}{4}(3.36d^2)L + \sin 30° \times \frac{1}{4}\left(\frac{3}{2}d^3N\right)\right]^2 + \left[1.2\left(\frac{3}{2}d^3N\right)\right]^2}.$$

Since, however, the extra weight of pulleys on a jackshaft is liable to produce a greater bending moment, it is customary to assume a larger shaft to transmit a given horse-power; therefore, instead of using $\dfrac{d^3N}{100}$ as the working horse-power transmissible by the shaft, it is better to use under these conditions $HP = \dfrac{d^3N}{125}$.

From this value of the power transmitted we obtain

$$F = \dfrac{4}{\pi}\phi\sqrt{\left[4\dfrac{\pi}{4}(3.36d^2)L + \sin 30°\times\dfrac{1}{4}\left(\dfrac{3}{2}\times.8d^3N\right)\right]^2 + \left[1.2\left(\dfrac{3}{2}\times.8d^3N\right)\right]^2}$$

$$= \sqrt{[0.8d^2L + 0.0114d^3N]^2 + [0.11d^3N]^2}.$$

The ratio of power absorbed by friction to the horse-power which the shaft is capable of safely transmitting will now become

$$\dfrac{HP_\circ}{HP} = \dfrac{0.0^s8dNF}{0.008d^3N} = \dfrac{0.10F}{d^2} \text{ per cent}, \quad . \quad . \quad (18)$$

from which Table XVII has been determined. In calculating the values given in this table it has been assumed that the belt speeds are the same as those previously considered, namely, 2640 feet per minute for main belt to head shaft, and one fourth of this, or 660 feet per minute, from head shaft to auxiliary shafting. This may be low in many cases, but, as already pointed out, the force F will decrease under the assumed conditions as the belt speed increases, so that we may expect the friction loss obtained by the above formula to be somewhat less than the actual.

In the foregoing discussion it has been assumed that the shaft transmits its allowable maximum power, that is, for line-shafting the power transmitted $= \dfrac{d^3N}{100}$, and for head-shafts the power transmitted $= \dfrac{d^3N}{125}$.

As a general thing, the actual average power transmitted by a shaft is not more than about three-fourths of its assumed working capacity; and since the weight and speed remain practically constant, the percentage of loss under conditions approximating those we have assumed will be somewhat greater than that given in the table. But while the power transmitted may be diminished 25 per cent, the percentage of increase in friction loss will vary between wide limits, depending upon the speed of rotation and length of shaft. Thus for a 3″ head-shaft 100 feet long, delivering three-fourths of its allowable capacity at 100 revolutions per minute, the loss increases from 9.0 to 11.5 per cent, which represents a gain of 28 per cent; at 250 revolutions the loss is now 14.0 per cent, corresponding to an increase of 14.3 per cent; while at 400 revolutions the loss is 17.8 per cent—a gain of only 10.6 per cent.

TABLE XVII.—POWER ABSORBED BY FRICTION IN JACK OR HEAD SHAFTS.

Diameter of Shaft in Inches.	Revolutions per Minute.	Percentage of Loss when Length in Feet = 25	50	100
2	100		4.8	8.5
	250		7.2	10.1
	400		10.0	12.5
2½	100		5.1	8.7
	250		8.3	11.0
	400		12.2	14.3
3	100		5.5	9
	250		9.5	12
	400		14.2	16.1
3½	100		5.8	9.1
	250		11.5	12.9
	400		16.6	17.8
4	100		6.3	9.5
	250		12.1	14.2
	400		18.4	20.5

For these determinations it is supposed that the shafting is properly supported, with hangers sufficiently close to each other to prevent undue deflection under working conditions, and that the shaft is in line, having good bearings lubricated as in common practice.* Departures from these

---

* In the above discussion it was assumed that the coefficient of friction is constant and that the friction varies directly as the load. While recent experiments on machine journals running in oil indicate that the coefficient of friction varies inversely with the load there seems no good reason to doubt the truth of Morin's laws for

assumptions will further increase the friction loss; but, on the other hand, this loss will be decreased if lighter or fewer pulleys be used throughout the length of the shaft, if the bearings be continuously lubricated, or if the machines be belted directly from the shaft below. Where shafting is employed there will generally be an additional loss due to the friction of the auxiliary shafting and counter-shafts, which is extremely variable.

It is outside the province of the present subject to discuss the losses in these secondary shafts: the losses which we have here been considering are those which exist in main line-shafting, jack- and head-shafts receiving their power presumably by leather-belting or ropes, with either of which for similar drives there should be no appreciable difference in the total weights of pulleys and shafting and the friction involved.

The diameter of grooved shaft-pulleys will be larger and the rim will be thicker; but for the same horse-power transmitted the width will be less and the weight not materially increased. In any case the total weight of grooved pulleys compared with that of belt-pulleys used in the same system is very small, and any individual differences may be neglected when taken as a whole. In those cases where ropes are used exclusively, as, for instance, in dynamo rooms and other power-stations, the pulleys are frequently heavier, and the shafting usually is fitted with a number of friction-clutches, thus materially increasing the weight on the bearings; in cotton-mills also, where the ropes are geared direct from the engine to the various floors of the mill, there is frequently a heavy stress on the shaft, especially on the upper floors, due to the weight of the ropes: under such

---

such comparatively rough and imperfectly lubricated bearings as we have been considering in which the friction between the rubbing surfaces in contact and not the viscosity of the lubricant is a measure of the resistance. See paper by Prof. Denton in *American Machinist*, Oct. 23, 1890.

conditions the shafting should be considered as head-shafts.

In work of this nature the velocity of the ropes is usually much greater both from the engine to the first shaft, and from the latter to the machine or secondary shaft. Assuming a speed of rope double that used for the previous tables, and letting the weight of pulleys, ropes, clutches, and couplings equal four times the weight of shaft, it can be shown that the formula for the friction load will be reduced to

$$F = \sqrt{[0.8d^2L \times 0.0057d^3N]^2 + (0.033d\,N)^2}.$$

TABLE XVIII.—POWER ABSORBED BY HEAD-SHAFTS CARRYING HIGH-SPEED ROPES.

Diameter of Shaft in inches.	Revolutions per Minute.	Percentage of Loss when Length of Shaft in Feet =	
		50	100
2¼	100	4.2	8.1
	250	4.8	8.6
	400	5.6	9.2
3	100	4.3	8.2
	250	5.2	8.7
	400	6.0	9.5
3¼	100	4.3	8.3
	250	5.3	8.8
	400	6.2	9.6
4	100	4.4	8.3
	250	5.6	9.1
	400	7.1	10.1

Line of action of resultant of tensions in main drive = 30°.
Velocity of ropes from engine 5280 feet per minute.
Ropes from shaft are horizontal, and run at an average of 2640 feet per minute.
All ropes drive from one side of the shaft toward the engine.
Weight on bearings four times weight of bare shaft.

From this formula Table XVIII has been calculated, and may be considered to represent the percentage of loss in the first shaft when working under full load; with lighter loads the percentage will be greater.

It must be noted that with any other arrangement of shafts the friction loss will vary. In the present case the ropes from the engine to the jack-shaft make an angle of 25° with the horizontal, and the ropes from the jack-shaft back to the machine or secondary shaft are horizontal, as in Fig. 54.

When rope-wells are used, each successively higher shaft will have an increased friction percentage, since the belt-pull becomes more nearly vertical, and the resultant load on each shaft is thereby increased.

It is worthy of remark that in long lines of shafting with high rim velocity the influence of belt-pull on the bearings is very slight compared to the weight of shaft and pulleys, so that the loss in friction is but little more than that due to weight alone. We see in this an additional argument for high rotative speeds in shafting, for, while the percentage of loss increases, from 8.1 to 9.2 in a $2\frac{1}{2}$-inch shaft 100 feet long running at 100 and 400 revolutions per minute respectively, the power transmitted by the shaft, as calculated from $HP = \dfrac{d^2 N}{125}$ increases from $21\frac{1}{2}$ for 100 revolutions per minute to about 87 h. p. for 400 revolutions per minute, so that while the friction percentage increases in the ratio $\dfrac{9.2}{8.1} = 1.14$, the power transmitted increases in the ratio of the number of revolutions per minute, or 4 to 1. In the first case the friction loss is $21\frac{1}{2} \times .081 = 1.74$ h. p.; and in the second, the loss is $87 \times .092 = 8.00$; therefore the net power transmitted by the shaft running at 100 revolutions per minute is 19 h. p. whereas by increasing the speed to 400 revolutions per minute the net power transmitted will be 79 h. p. With higher belt velocities and increased rotative speeds in our factory shafting the friction loss, instead of being from 30 to 50 per cent of the total power transmitted, ought not to exceed one-half of these percentages; for with higher speeds narrower and lighter pulleys could be used, the belts could be run slacker, and lighter shafting could be employed.

Although it has been previously considered in a general

way that the friction due to the shafting may be taken as about 10 per cent of the full load transmitted to the shaft, yet, in the light of further investigation, it will be seen that, owing to the various conditions under which the shafting is run no general value for friction loss can be assigned.

While the friction absorbed by large engines is reasonably less for rope-geared fly-wheels when compared with engines using flat belts, driving in a similar manner, there seems to be no good reason for supposing that the friction in ordinary mill-shafting should be appreciably different when rope-driven.

On account of the larger diameter of pulleys used with rope-driving, the velocity of the rope may be, and usually is, greater than that in a belt used in the same place, and for this reason the pull on the shaft due to the tensions in the rope may be less; but with long lines of rope-driven mill-shafting the main drives only are of rope, and any difference in pull on the bearings which might exist in favor of the rope-driven main shaft must necessarily be small when compared with the total friction load due to the pull of the numerous cross or machine belts which, running at a greatly reduced speed, produce by far the greater effect on the shaft. With short lines of shafting, however, there will generally be a small saving in favor of a rope-driven plant. Under these conditions the effect of the ropes to and from the first motion shaft is usually in excess of that due to the belts which may be used to transmit power from the main or jack shaft to secondary shafts or machines; and therefore, since the rope-pull is less than would be produced by belts used in the same place, we may expect the friction to be less. When ropes are used entirely, as in electric and other power stations, we should expect the friction loss to be somewhat less, assuming that the ropes are run at a higher speed than would be used for belts in the same place.

CHAPTER XI.

It has been stated that a further decrease of from 5 to 8 per cent of the power transmitted by a rope may be attributed to losses in the rope itself due to resistance to bending, wedging in the groove, differential driving effect, and creep, all of which affect the loss to a greater or lesser extent.

Various formulas have been proposed by several eminent authorities by which the resistance of a rope to bending might be determined. Eytelwein's formula assumes that the resistance of a rope is directly proportional to the tension and the square of the diameter, and inversely proportional to the radius of curvature of the pulley; in which case the stiffness of a hemp rope for each winding and unwinding is given by

$$\sigma = c\frac{d^2}{r}T,$$

where c is a constant equal to 0.23, d is the diameter of rope in inches, r is the radius of pulley in inches, and T is the tension in the rope. If the ratio of the diameter of rope to the diameter of pulley over which it runs equals 1 to 30, the above formula becomes, for a rope running over two pulleys,

$$\sigma = 0.03dT.$$

Reuleaux states that, since transmission-ropes are usually quite slack, the coefficient of stiffness should be taken somewhat less than Eytelwein's value, and suggests that two-thirds would represent a fair approximation; this would give

$$\sigma = \frac{2}{3} \times 0.23 \frac{d^2}{r} T = 0.15 \frac{d^2}{r} T$$

for each pulley in the system.

If the tension on the tight and slack sides of the rope be represented by T_1 and T_2, respectively, the average load on the rope may be considered equal to $\frac{1}{2}(T_1 + T_2)$; if, further, the conditions be assumed such that the slack-side tension equals one half that in the tight or driving side,— we shall have $T = \frac{1}{2}(\frac{3}{2} T_1)$. Hence for two pulleys, when the diameter of the latter equals 30 times the diameter of rope,

$$\sigma = 2 \left(\frac{0.15 d^2}{15 d} \times \frac{1}{2} \times \frac{3}{2} T_1 \right)$$
$$= 0.02 d \times \tfrac{3}{4} T_1.$$

Since T_1 has been taken in our previous work as equal to $200 d^2$ pounds, the stiffness in the rope will now become

$$\sigma = 0.02 d \times 150 d^2$$
$$= 3 d^3.$$

Now under these relations of tension the driving force may be obtained from

$$P = T_1 - T_2 = 200 d^2 - \frac{200 d^2}{2} = 100 d^2;$$

hence the ratio of loss due to bending will be

$$\frac{\sigma}{P} = \frac{3 d^3}{100 d^2}.$$

For a ¾-inch rope running over two pulleys the loss equals 2.25 per cent, while for a 2-inch rope under similar conditions the loss becomes 6 per cent. This is what we might expect; for it is reasonable to suppose that the percentage of loss should increase with the diameter of rope.

It will be noticed that the work done in bending a rope over its pulley is directly proportional to the number of bends, and therefore in designing a rope transmission every effort should be made to restrict the number of bends, as this is not only a large factor in the wear, but, as just shown, the power transmitted for a given tension is constantly reduced as the number of bends increases.

This feature is a decided disadvantage with installations on the continuous-rope system where one rope is bent in both directions around a number of pulleys on the several floors of a factory. Under such conditions, and also in those special cases where it is absolutely necessary to run pulleys smaller than that obtained from the formula* $D = d^{1.7} \times \sqrt{V+12''}$, the user should put in the very best quality of loosely twisted rope and run it at a less strain than would otherwise be adopted; for under such conditions the flexibility and elasticity of the rope are more desirable than a high breaking strength.

While the foregoing formula of Eytelwein may give a measure of the force required to bend a rope over a pulley under a certain set of conditions, it will be evident, since the conditions vary considerably in different installations, that, in order to be generally applicable to any given case, a formula must contain other factors than those included in Eytelwein's and other similar formulæ. In a flying rope running in V grooves, in addition to the bending of the fibres there is a permanent reduction in cross-section, due to the uniform compression of the rope, as will be noticed by measuring a rope that has been running some time; besides this there is a temporary deformation due to the distortion of the rope as it passes over the pulley.

---

* See page 179.

The flexibility and elasticity of the rope undoubtedly have much to do with the resistance to bending, and the degree of twist put into a rope is an important factor in this connection; for although hard-twisted ropes are very much stronger than those loosely twisted, the internal wear is much greater, as the fibres are held more rigidly and do not slide as freely upon each other. The advantages of lubricating the fibres of a manilla rope have been already discussed; it is sufficient to state here that the degree of lubrication affects the flexibility of the rope, and hence enters as a factor in determining the loss due to bending. The varying angle of contact and, to a lesser extent, the angle of the groove, must also have a certain influence upon the resistance. In view of these considerations it will be seen that any deductions from existing formulæ are of doubtful utility when applied to transmission-ropes in use, and should be considered only as relative, and not absolute.

Another source of loss is that due to the wedging of the rope in the groove. Although this action exists to a greater or less extent in all rope transmissions where the shape of the groove is such that the rope does not bottom, yet it is undoubtedly true that its effect in a well-constructed plant has generally been over-estimated. That it does exist to a harmful degree can be seen in many installations from the way in which the tight side of the rope follows upon the driven pulley. With the single-rope method this is especially true in new installations, where the tension is purposely made rather high to allow for stretch in the rope; it is, however, frequently found in the continuous-rope system where the tension-carriage is over-weighted. The factors which enter into the consideration of this loss are: Tension on the driving and slack sides of the rope, the angle of groove, and the velocity, weight, and condition of the ropes. As we have previously shown (page 132), the driving-side tension T_1 is made up of three parts, namely, the driving force P, the centrifugal force F_0, and the tension T_s, necessary to balance the strain for adhesion; that is,

$$T_1 = P + F_0 + T_s.$$

In like manner the tension in the slack side of the rope consists of the strain necessary for adhesion plus the strain due to centrifugal force, that is,

$$T'_2 = T_s + F_0.$$

It has been claimed that no loss can occur in pulling the rope out of the groove, since the centrifugal force set up in the rope is many times greater than any caused by the tension on the slack side when leaving the pulley; but it is obvious that a part cannot be greater than the whole, and therefore the centrifugal force, while greatly reducing the wedging force, cannot altogether eliminate it.

An abnormal degree of slack-side tension has a direct effect upon the wedging of the rope in the groove, for if sufficient sag is not allowed and the slack-side tension is, therefore, needlessly great, it follows that the power transmitted will be reduced, since the driving force, P, is equal to the difference of the tension in the two portions of the rope $= T_1 - T_2$; on the other hand, the force drawing the rope into the groove will be increased, since the force equals

$$T_1 - F_0 + T_2 - F_0^* = P + T_s + T_s$$
$$= P + 2T_s.$$

It is obvious, therefore, that the slack-side tension

---

* Although the centrifugal force increases the tension in the rope, its tendency is to cause the latter to leave the pulley, and therefore it should not be considered as a part of the forces drawing the rope into the groove.

should be no greater than just sufficient to give adhesion to the ropes and prevent undue slip at the desired speed.

As to the best angle for the groove in the pulley, opinion is still somewhat divided; but in England the general practice seems to favor an angle of 45° as the most suitable.*

In the earlier installations a more acute angle was used, as evidenced by the discussion of Mr. Durie's paper on Rope-driving, presented to the British Institution of Mechanical Engineers.† Grooves having an angle of 30° had been tried, but it was found that the wear on the rope was altogether too great; 40° was a very satisfactory angle, and is still preferred by many engineers, while others use as large an angle as 60°. Occasionally half-round grooves are used; but with semicircular grooves on cast-iron pulleys either the tension in the rope must be increased or a greater number of ropes must be employed: in any case the advantage seems doubtful. With wooden-rimmed pulleys, however, the semicircular groove is the better form; for since the coefficient of friction on the wooden pulley is from thirty to fifty per cent greater than for a similarly shaped groove on an iron pulley, it follows that the tension in the rope need not be so great; moreover, wooden pulleys that have been in use for some time would indicate that the semicircular groove is better adapted to the work, for with anything but very light loads the angular groove is soon cut out by the rope, producing a rim somewhat similar to the one shown in Fig. 57, which represents an angular-grooved wooden pulley that had been in use but a few months.

With a semicircular groove in the first place the latter will retain more nearly its original form, and the wear on the rope will be greatly reduced.

The pliability of the rope has considerable influence on

---

* M. E. in *American Machinist*, Nov. 10, 1892.
† See Proc. Inst. M. E. 1876.

the shape of the groove; while a 30° angle may give the correct shape for a soft, loosely twisted cotton-rope, a harder twist may require an angle of 40° or even 50°; in the same way a 40° groove may be all right for some makes of manilla rope, while others of a less yielding nature would give better results if an angle of 50° or even 60° were used.

At the present time there are very few pulleys used in this country, except for machine bands, having grooves with an angle less than 40°. Formerly the Yale & Towne

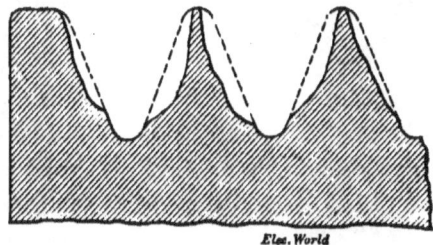

FIG. 57.—RIM OF WOODEN PULLEY SHOWING WEAR.

Mfg. Co. employed grooves of 30° for small cotton ropes (about ¾ inch in diameter) driving their travelling-cranes; but these were subsequently changed to 50°, and at the same time larger ropes, 1¼ inch, were adopted.

For manilla rope-drives one large manufacturer uses an angle of 60° on all his pulleys, whether of wood or iron; this angle was arrived at after much trial, and represents an experience with manilla-rope transmission, covering a great many years.

The most suitable angle of groove is that which affords the greatest frictional adhesion without undue slip, and at the same time offers the least resistance to the rope in leaving the groove: with much slip the rope is rapidly worn out, while with an excessive grip the wear is also rapid, and a relatively large amount of force is absorbed in overcoming the wedging. The usual practice in this country for both cotton and manilla ropes is an angle of 45°; in many cases the section of the groove is formed by arcs of circles, having a radius equal to from 3 to 4 diameters of rope, in which case the included angle is constantly changing, and the coefficient of friction, and hence the grip, will vary with the diameter of rope used. Thus in Fig. 58, if AB and $A'B'$ represent two ropes of slightly different diameters running over a grooved pulley, the one which sinks deeper into the groove will include the greater angle of contact ABC, and hence the lesser coefficient of friction. This will hold true of two ropes having the same diameter but different degrees of twist: the harder rope will not sink as deeply into the groove, and its coefficient of friction will be greater than that of the smaller rope, other things being equal, on account of the lesser angle in the groove at the point of tangency of the harder rope.

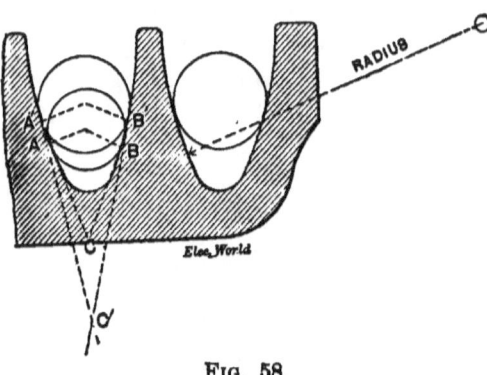

FIG. 58.

With the excellent ropes that are now being made for transmission purposes this form of groove possesses many advantages, even with the continuous-wound system.

In this system the tension is assumed to be the same in each wrap; but there is unquestionably a variation of pull due to momentary fluctuations, which must either be absorbed by the elasticity of the rope or transmitted through the rope until the strains are equalized. If the rope used is uniform in structure, that portion which receives the greater strain will be drawn more deeply into the groove; and if the latter be of the curved form, the coefficient of friction will be reduced: so that the resultant adhesion produced in this portion of the rope will be less than that in some other part which, under a lighter strain, occupies a position on the pulley such that its coefficient of friction, and consequent adhesion, is greater for a given back-tension.

With the usual arrangement of pulleys, namely, that in which the larger wheel is the driver, the tendency of an increase in tension is to increase the velocity of the driven pulley. If several wraps of a series occupy positions in the grooves such that with a given ratio of pulley diameters the velocity of the smaller pulley is twice that of the driver, any decrement, x, of the effective radii of another wrap, which is drawn more deeply into the grooves of the two pulleys, will alter the velocity ratio from $\frac{2}{1}$ to $\frac{2-x}{1-x}$, a quantity which must of necessity be greater than 2. The tendency of this wrap then is to produce a greater velocity in the driven pulley, which cannot occur without some slip in the other wraps; but these wraps have a greater adhesion, and therefore tend to drive the pulley at a less number of turns per minute, which will produce slip in the more heavily strained member: hence the effect of any change in position due to sudden increased strain on one wrap will tend to quickly adjust the tensions in all positions of the rope and neutralize any inequalities in driving effort. This form of groove, as we shall show subsequently, is even more desirable for the individual rope system, where the evils of differential driving are frequently so pronounced.

As noted previously, the frictional grip depends both upon the arc of contact of the rope with the pulley and the coefficient of friction, which latter varies with the angle of the groove. In order, therefore, to produce the same friction on each pulley, **the product of** the arcs of contact by their respective coefficients of friction must be equal. If, as before, we take the coefficient of friction of a well-lubricated rope on a smooth, flat metal pulley equal to 0.12, the coefficient for the same rope in a groove whose angle is θ degrees will be $\phi = 0.12 \operatorname{cosec} \dfrac{\theta}{2}$; hence with arcs of contact α and α' we should have for an equal grip on each pulley

$$\phi\alpha = \phi'\alpha',$$
$$0.12 \operatorname{cosec} \frac{\theta}{2} \times \alpha = 0.12 \operatorname{cosec} \frac{\theta'}{2} \times \alpha',$$

assuming that the multiplier 0.12 will give the correct coefficient of friction on each pulley, since the percentage of slip is to be the same.

As the numerical value of the cosecant of an angle varies inversely with the angle, it will be obvious that the pulley, having the lesser arc of contact, should also have the more acute angle in the groove. This property of rope friction is frequently taken advantage of in designing a plant, and it is not uncommon to find the grooves in the large wheel more obtuse than those in the smaller, especially when there is considerable difference in the diameters of the pulleys.*

From the above equation there is obtained

$$\operatorname{cosec} \frac{\theta}{2} = \operatorname{cosec} \frac{\theta'}{2} \times \frac{\alpha'}{\alpha},$$

in which θ and α represent the angle of groove and arc of contact, respectively, on the larger pulley, and θ' and α' similar values for the smaller pulley. With the least angle of groove equal to 35°, 40°, or 45°, the corresponding angle in the larger pulley should be as indicated in Table XIX, when the ratio of the arcs of contact is known.

---

*Mr. T. Spencer Miller, of the Lidgerwood Mfg. Co., has been granted a patent (U. S. patent No. 444919), upon the application of this principle to continuous-rope transmissions, in which with sheaves of different diameters more obtuse grooves are given to a larger wheel.

TABLE XIX.—ANGLE OF GROOVE FOR EQUAL ADHESION.

Arc of contact on small pulley / Arc of contact on large pulley $= \dfrac{a'}{a}$	0.9	0.8	0.75	0.7	0.65	0.6
Angle of groove in large pulley when groove in small pulley = 35°	40°	44°	74°	51°	55°	60°
Angle of groove in large pulley when groove in small pulley = 40°	45°	50°	54°	58°	64°	70°
Angle of groove in large pulley when groove in small pulley = 45°	50°	55°	60°	66°	72°	80°

Of course an idler or binder pulley may be used to increase the arc of contact on the smaller pulley, and thus maintain an equality of grip on each pulley; but this device produces an objectionable reverse bend in the rope, which should be avoided as much as possible in rope transmission. Both of these arrangements, as well as the winder-pulley previously discussed, are intended to prevent slip and at the same time obtain the maximum adhesion for the least amount of back-tension without increasing the losses due to wedging in the grooves, journal-friction, and wear in the rope.

As pointed out by Mr. W. H. Booth,[*] although there may be some loss and wear due to wedging in the groove, by far the greater loss is occasioned by the deeper wedging of one rope as compared with another, so causing them to grip upon a different circumference, in which case each rope tends to impart a different velocity to the driven pulley; the actual resultant velocity will be a mean of the several velocities of the individual ropes, so that slipping and wear of some or all of the ropes must occur, due to the differential driving thus set up.

With continuous rope transmissions this effect is not so apparent, although it exists to a certain extent on account of inequalities in the rope and various mechanical imperfections in the system; but in the English or independent

---

[*] *American Machinist*, Dec. 8, 1888.

rope transmission its effect is very marked, and is generally considered as the principal source of loss of efficiency in this system.*

While there is undoubtedly a considerable loss which may be charged to this cause, it is also true that its effect may be greatly reduced by a careful study of the requirements of the problem and an intelligent application of correct principles to the case in hand.

It is evident that in order to prevent slip and the loss of power incident thereto it will be necessary to obtain a uniform velocity in the several ropes running over a pair of pulleys; to approach this desideratum each set of ropes should be of the same make and degree of hardness, of uniform diameter, evenly spliced, having the same amount of sag in both members, and run over grooves of uniform diameter, shape, and smoothness.

Obviously it would be impossible in practice to maintain all these conditions, even if it were practicable to overcome the mechanical difficulties and install a plant under the given requirements; yet much may be done to effect the desired end.

Uniformity of length is an important requirement, for, if one rope of a set be allowed a less amount of sag than the others, the sum of the tensions will be greater in this rope, and in consequence it will be drawn more deeply into the groove; its pitch diameter will therefore be reduced and its velocity will be different from the others in the set, in which case if the driving and driven pulleys are of unequal diameters the tendency will be to give the driven pulley a different velocity, and slip must necessarily occur.

The following from the *American Machinist* is pertinent to the subject:

"It will be readily seen that in a set of ten or a dozen

---

* *American Machinist*, Dec. 1, 1892.

ropes, each of a somewhat different length, the loss of power from this cause may easily become a very serious item, and to this there is to be added the correspondingly diminished life of the ropes. A similar action occurs when worn ropes are allowed to work conjointly with new, even though the deflections, and therefore the tensions, on the several ropes are practically equal. In this case the loss due to abnormal tension and wedging of some of the ropes into the grooves is avoided, but the differential driving effect due to the ropes virtually running on pulleys of different diameters still exists, and is equally objectionable. It may here be noted that the effect of the differential driving upon the ropes depends to a very large extent upon the relative diameter of the two pulleys.

"It is evident that, when the driving and driven pulleys are of the same diameter, any variation in the effective pitch diameters of the several grooves will have no appreciable effect upon the transmission, provided that the diameter and shape of the corresponding grooves in the two pulleys are the same. It may be noted, however, that the worn ropes which run deeper in the grooves, having a slightly less velocity, are subjected to a somewhat greater stress than their newer and larger companions."

With a form of groove similar to that shown in Fig. 58, in which the sides of the groove are circular arcs of large radius, there will be a tendency to correct the differential driving, especially so when the driving pulley is smaller than the driven. In this case, if we assume that all the ropes in a set are put on with the same amount of sag, so that the tensions are practically equal for both old and new ropes, any difference in diameter of rope will cause the larger to carry more than its share of the load, since the effective radius of the pulley, and therefore the velocity of the rope, is greater; moreover, since the coefficient of friction varies inversely with the depth of contact in this form of groove, the larger rope is acted upon by a more intense grip, by reason of which more work is imposed upon it, while a smaller rope will be relieved of some of its work.

The tendency of the larger rope is to turn the driven pulley at an increased velocity. Thus in Fig. 59 if D be

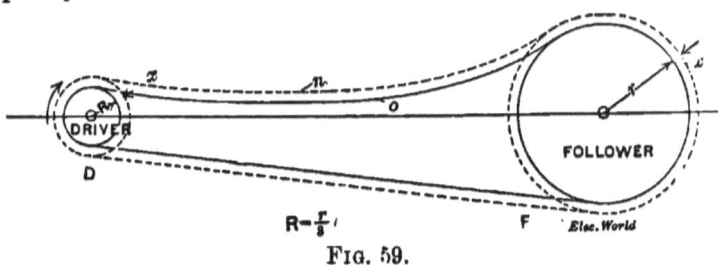

FIG. 59.

the driving-pulley and F the follower, with respective diameters such that the velocity ratio $\dfrac{R}{r}$ is 1 to 3, then the smaller ropes o will tend to drive F in the ratio $\tfrac{1}{3}$, while the larger ropes, n working at a distance x further away from the centre of pulleys, will tend to drive F in the ratio $\dfrac{1+x}{3+x}$, and therefore at a higher velocity. The resulting velocity of the follower will depend upon the work done by each rope, so that some slip must necessarily occur; on account of the greater load taken by the larger ropes they will be rapidly compressed and worn: hence any initial variation in turning effort will be speedily reduced to a minimum by this equalizing process, which, although it incurs loss, ultimately insures a better distribution of the work with the least wear on the ropes.

On the other hand, when the driving-pulley is larger than the follower the smaller ropes are drawn further into the grooves and tend to impart an increased velocity to the driven sheave. Thus in Fig. 60 the large pulley is the driver and the velocity ratio is $\dfrac{R}{r} = \dfrac{3}{1}$ with the smaller ropes o; with the large ropes n, working at a distance x nearer the circumference, the tendency will be to produce a ratio equal to $\dfrac{3+x}{1+x}$: hence the effect will be retardation. Since the larger ropes acting higher in the groove have an increased grip and speed they will exert a greater influence upon the smaller pulley; and although the smaller ropes may be drawn more deeply into their grooves in attempt-

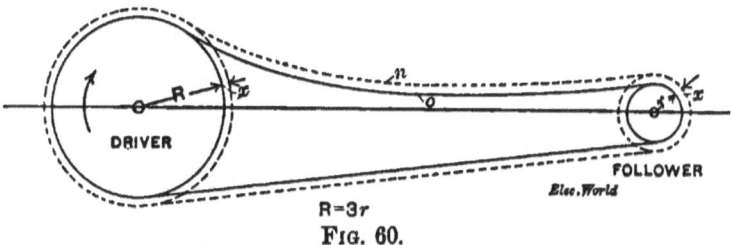

FIG. 60.

ing to drive the follower at a greater number of revolutions per minute their effect is lessened on account of the diminished grip, due to the form of groove, in consequence of which the larger rope acts with a greater effect and the smaller with a lesser effect than would obtain with an ordinary V-groove under like conditions.

To equalize the driving efforts of a number of ropes, and to prevent the slip which must inevitably occur with a solid-grooved rim, Mr. John Walker has devised and patented * a "differential" driving-pulley, since it allows the ropes to travel at different speeds suited to the conditions imposed upon each rope.

Originally intended for cable-railway machinery, where the wear on the drums due to the wire cable is excessive, the differential principle has been extended to other uses, notably elevator sheaves and rope-transmission pulleys.

In the latter the rope is led over a number of separate

---

* Feb. 23, 1892.

rings, Fig. 61, adapted to turn loosely and independently of each other on the smooth circumference of the drum.

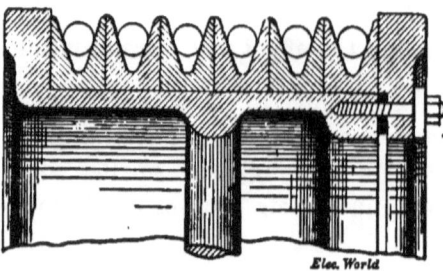

FIG. 61.

While the rope is passing over the pulley the tendency of the rings will be to adjust themselves to the strain in each member by moving around the circumference of the drum. Thus the driving-tension is equalized, and each rope is brought to do its own share of the work without slipping in its groove. These rings have a diametrical friction, due to the pressure of the rope in the groove transferred to the flat surface of the drum. In addition to this an adjustable rubber washer is inserted between the rim of pulley and loose flange, so that by tightening this adjustable joint the separate rings are caused to exert enough pressure upon each other to produce a certain amount of friction on the side surfaces, the combined friction of the several rings being sufficient to drive the pulley or ropes, as the case may be. In this way each rope bears its due share of the work, as the adjustment is such that the friction between the several parts is brought into equilibrium.

In practice the axial rotation of the ropes will frequently exert a modifying influence on the differential driving, since, other conditions being unchanged, a rotating rope tends to maintain its circular form, and therefore will work less deeply into the groove. An additional advantage is that such rotation promotes the durability of the rope, as the wear is more uniform.

The loss due to the elastic slip or creep of the belt has some influence upon the efficiency of transmission, but in any case its effect is small. When an elastic body, such as a rope, is placed under tension, it stretches, and the elongation, within the limit of elasticity, is proportional to the strain in the rope. When power is transmitted from one pulley to another the driving side is subjected to a greater tension than the slack side, in consequence of which the velocity in the driving side will be slightly greater than that in the slack side, and the circumferential velocities of the two pulleys will not be the same. This will be evidenced from the following considerations:

Let V = circumferential velocity of driver;
V' = " " " follower;
T_1 = tension in driving side of rope;
T_2 = " " slack side of rope;
A = cross-section of rope in square inches;
L = original length of a piece of the rope in either
    = side in its normal condition;
e = elongation in driving side due to tension T_1;
e' = " " slack side due to tension T_2;
E = modulus of elasticity of the rope.

When the rope is at work $e = \dfrac{T_1}{EA}$ and $e' = \dfrac{T_2}{EA}$, so that the length of each member will now be $L + e$ and $L + e'$, respectively. The length of rope running on to the driving-pulley in a unit of time will therefore be greater than that delivered to the driven pulley in the proportion $\dfrac{L+e}{L+e'}$, and the velocity of the two pulleys will now be

$$\frac{V}{V'} = \frac{L+e}{L+e'}; \quad \text{hence } V' = V \times \frac{L+e'}{L+e}.$$

Calling L unity, and assuming $\dfrac{T_1}{T_2} = 2$, $\dfrac{T_1}{A} = 320$ pounds (since $T_1 = 200d^2$ pounds, and $A = 0.8 \times \dfrac{\pi}{4}d^2$), and $E =$

40000,* we have

$$e = \frac{320}{40000} \quad \text{and} \quad e' = \frac{320}{2 \times 40000};$$

therefore

$$V' = V \times \frac{1 + \frac{320}{80000}}{1 + \frac{320}{40000}} = 0.996\,V;$$

that is, the loss due to creep when all the ropes are working under normal conditions is about one half of one per cent. With different ratios of driving to tight-side tension and different intensities of stress it is evident that this loss will vary in the different ropes; in any case the loss ought not to exceed one and a half per cent, as will be seen if we assume extreme conditions.

For instance, let the stress on the rope be 600 pounds per square inch of section and assume the ratio of tensions equals one to five; with the most elastic long-staple cotton rope it is possible that the modulus of elasticity may be reduced to 30,000 pounds; therefore, under these conditions, the velocity of the circumference of the driven pulley will be

$$V' = V \times \frac{1 + \frac{600}{5 \times 30000}}{1 + \frac{600}{30000}} = 0.984\,V,$$

thus representing a loss of 1.6 per cent. It is probable that one per cent is an ample allowance, even under unfavorable conditions.

---

* See Reports of Tests on Manilla Rope, Ex. Doc. No. 36, 1885.

CHAPTER XII.

The construction of rope pulleys is a matter of considerable importance, for it is evident from what has preceded that the size, shape, and condition of the pulley all exert a marked influence upon the efficiency of rope transmission.

It has been shown that the size of the pulley materially affects the wear of the rope: the larger the sheaves the less the fibres of the rope are flexed, and the less they slide on each other; consequently there is less internal wear of the rope.

The minimum diameter of the pulley, as given by different authorities, varies from thirty to forty times the diameter of rope to be used. Forty times the diameter of rope for manilla is excellent practice, and experience has shown that a larger multiplier would be still better, as the larger a pulley, the better for either belt or rope passing over it; but such a rule, although convenient to use, was evidently founded upon large-sized ropes running at a high velocity, and a little consideration will show that while forty times the diameter of rope may be all right for a two-inch rope, it is also true that a lesser proportion will give suitable diameters of pulley for a one inch rope. If we take two ropes d_1 and d_2, whose diameters are one and two inches, respectively, and bend them over pulleys of the same diameter, the fibres in d_1 will be stressed an amount equal to x due to the stretch and sliding of the fibres one upon another. On the other hand, since the outer fibres in d_2 are twice as far from the neutral axis, these fibres will stretch and slide upon each other to an extent equal to twice that produced in the smaller rope; but the relation of the areas of the two ropes is

$$\frac{\frac{\pi}{4}d_2^2}{\frac{\pi}{4}d_1^2} = \frac{d_2^2}{d_1^2} = 4$$

under the given conditions; hence the total slip in the larger rope, due to the sliding of the fibres upon each other, will be eight times greater than in the smaller one, or as $\frac{d_2^3}{d_1^3}$. Therefore, for equal extension and slip of the fibres in bending a rope over a stationary pulley, it would appear that the pulleys should have diameters proportional to the cube of the diameters of the ropes.

In the above consideration no account has been taken of the external wear, produced by slipping in the groove, wedging, rubbing contact, and other causes of loss which affect a running rope; for a given velocity these losses will increase with an increased speed of rotation, and hence will be greater with a smaller pulley, which would indicate that the diameter of the latter should not be directly proportional to the cube of the diameter of the rope. It is also evident that a rope running at 5000 feet per minute will be subjected to a greater number of bends and a greater external wear than it would if running over the same pulleys at 2000 feet per minute. The first of these influences is recognized by many engineers who use a rule approximating the following:

For the least diameter of pulley, D, multiply the circumference of rope by ten times its diameter and divide by two.* This is practically equal to $D = 15d^2$ inches.

While we believe in using as large a pulley as possible in

---

*Mr. Jas. Gamble in *Textile Recorder*.

any given case, conditions will arise when it is desirable to use the smallest possible diameter without excessive injury to the rope. In such case the individual circumstances should be considered by the designer.

From the foregoing it is obvious that for a given rope and tension the least diameter of pulley which may satisfactorily be used under known conditions should be dependent upon the velocity of the rope, and should vary with the size of the latter in such a manner that for any given speed the pulley diameter will be proportional to some power, greater than unity and less than the cube of the diameter of the rope.

From an investigation of numerous examples in operation under varying conditions, some of which work satisfactorily and others very poorly, it would seem that while the value $15d^2$ may give a suitable diameter of pulley for a soft cotton rope of small diameter, such a value is entirely too small for an equal-sized manilla rope; but, on the other hand, $40d$ is somewhat larger than absolutely necessary for these ropes, although it is always desirable to use such a pulley if conditions permit.

If for any reason it is necessary to adopt a small pulley, the least pitch diameter for a sheave to be used with manilla rope working under an assumed tension of $200d^2$ pounds may be determined from the following empirical formula, which is believed to represent the requirements of good practice:

$$D = d^{1.7} \times \sqrt[4]{V} + 12'',$$

in which D = pitch diameter of pulley in inches;

d = diameter of rope in inches;

V = velocity of rope in feet per minute.

In order to simplify the use of this formula the following values of $d^{1.7}$ and $\sqrt[4]{V}$ have been calculated as given in Table XX.

Table XXI gives values of D for ropes varying from $\frac{3}{4}$ inch to 2 inches in diameter, running at 2000 to 5000 feet per minute. When the speed of the rope is not known the diameter given in the last column should be used for the minimum size of pulley; in fact it would be better to use these diameters, or even larger ones, in all cases, provided the constructive features in the plant will permit their application. With an increased tension in the ropes the diameter of pulley should be increased also. In the same way if the working tension should be less than $200d^2$ pounds, then the diameter of pulley may be less than here given. If cotton rope be used, the least diameter of pulley may also be taken somewhat less.

TABLE XX.—VALUES OF $d^{1.7}$ AND $\sqrt[3]{V}$.

Dia. of rope d	$\frac{3}{4}$	1	$1\frac{1}{4}$	$1\frac{1}{2}$	$1\frac{3}{4}$	2
Value of $d^{1.7}$	0.61	1	1.46	1.99	2.59	3.25
Velocity of rope, V, in feet per minute	1000	1500	2000	3000	4000	5000
Value of $\sqrt[3]{V}$	10	11.4	12.6	14.4	15.87	17.1

TABLE XXI.—LEAST DIAMETER OF PULLEY FOR GIVEN DIAMETER AND SPEED OF MANILLA ROPE.

$$D = (d^{1.7}) \times \sqrt[3]{V} + 12''.$$

Diameter of Rope.	Velocity of Rope in Feet per Minute.			
	2000	3000	4000	5000
$\frac{3}{4}$	20	21	22	$22\frac{1}{2}$
1	$25\frac{1}{2}$	$26\frac{1}{2}$	28	29
$1\frac{1}{4}$	$30\frac{1}{2}$	33	35	37
$1\frac{1}{2}$	37	$40\frac{1}{2}$	$43\frac{1}{2}$	46
$1\frac{3}{4}$	44	49	$53\frac{1}{2}$	$56\frac{1}{2}$
2	53	$58\frac{1}{2}$	$63\frac{1}{2}$	67

ROPE-DRIVING. 181

TABLE XXII.—ROPE PULLEYS FOR GENERAL WORK.

Diameter of Rope..	¾	1	1¼	1½	1¾	2
Diameter of Pulley	24	36	48	60	72	84

It is well to remember that the cause of many failures and much trouble experienced in rope-driving is due to the use of too small a pulley for the size of rope and tension carried, but we have yet to hear of a case where the diameter of pulley has been too great. When it is not absolutely necessary to restrict the diameter to the smallest possible which may be used, the least diameter should conform to that given in Table XXII, which has been arranged for general work, and gives least diameters, to the nearest half foot, for rope-pulleys suitable for all speeds within the limits of good practice.

When a close velocity ratio between driver and follower, greater or less than unity, is required, the pitch diameter of each pulley should be measured from the point of tangency of the rope in the groove, and not from the centre of the rope. For, if D represent the diameter of driver measured to centre of rope (Fig. 62), and F the corresponding diameter of follower, the velocity ratio will be $\dfrac{D - 2x}{F - 2x}$, where x is the common vertical distance between centre of rope and point of contact. Where the pulleys are the same size it is obvious that the velocity ratio will not be changed, but as the difference in diameter increases the influence of x will be more marked; thus if $D = 3F$, the velocity ratio will be $\dfrac{3F - 2x}{F - 2x}$,—a result manifestly greater than $\dfrac{3F}{F}$. In any case, the smaller the diameter of pulleys for a given velocity ratio, the greater the effect of the quantity x.

In ordinary single transmissions it will be sufficiently close to assume the diameter of pulley as measured from centre of rope.

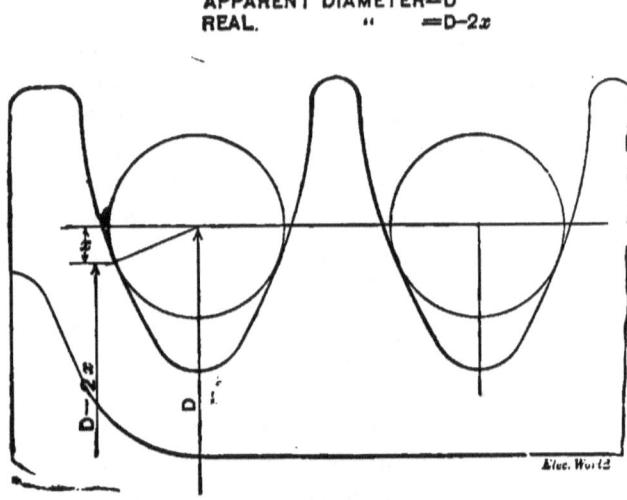

APPARENT DIAMETER=D
REAL.   "   =$D-2x$

FIG. 62.—DIAMETER OF PULLEY.

It is evident from the previous considerations that the condition and shape of the groove is a matter of much importance: so generally is this recognized, that manufacturers now almost universally turn their rims to special gauges and templets, which insure uniformity in diameter and shape of groove. Special tools have been devised for turning the grooves, and one prominent manufacturer has constructed a special machine in which the rims are milled out. By this process some of the smaller pulleys are machined in one operation—as many cutters being used as there are grooves to be milled. Uniformity of pitch, diameter, and contour are thus insured independently of the operator. Formerly the bottom of the groove was furnished with spikes, or the sides were cut into angular teeth in order to prevent the rope from slipping;* but a greater experience with rope-driving has shown that the whole surface of the groove must be perfectly smooth, and should be carefully polished as well as machined, since the fibres of the rope, if allowed to rub on a rough-turned or cast surface, will gradually break, fibre by fibre, and thus give the rope a short life. It is also necessary to avoid using any pulley with sand or blow holes in the groove, as they are very destructive to a rope: when blow-holes occur, if not honeycombed excessively, they should be filled in with lead, or, preferably, Babbitt's metal; otherwise the pulley should not be used.

Some rope-pulleys are simply cast and the rims smoothed up by holding a piece of abrasive material, such as a broken emery-wheel or grindstone, against the surface of the groove while the pulley revolves at a high speed. While this may produce a smooth surface, such an expedient is doubtful economy; for no matter how carefully a multiple-grooved pulley may be cast, it is almost impossible to obtain a rim all the grooves of which are true and of the same diameter: the result is that, with such pulleys, the ropes tend to vibrate and sway from side to side, rubbing against each other, and frequently against side-posts, walls, floors, and other obstructions, which rapidly destroy the rope.

Attempts have been made to produce a finished surface in the groove by casting the rim in an accurately turned chill, but such pulleys have not been as successful as anticipated. A form of pulley made by the Link-Belt Engineering Co. of Philadelphia consists of an iron sheave cast in rings, some with and some without arms and hub, from which a complete pulley is built up having the requisite number of arms for strength, and at the same time being

---

* Willis: "Principles of Mechanism."

light and free from excessive shrinkage strains liable to exist in wheels having light arms and heavy rims and hubs. The sheaves have a slight projection on one side and corresponding recess on the other, with bolt-holes at the circumference, so that a multiple-grooved pulley with any number of grooves may be readily built up by bolting the sections together as shown

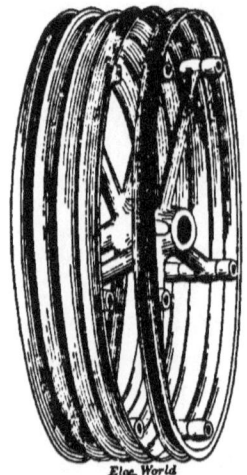
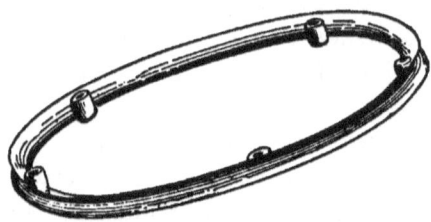

FIG. 63.—BUILT-UP ROPE PULLEY FOR LIGHT WORK.

in Fig. 63. These wheels are made from metal patterns and moulded in a three-part iron flask. The core which makes the groove is continuous and of green sand, producing a very smooth casting. The groove is finished with an emery-wheel swinging in a frame like a cut-off saw, the final finish being given by the use of emery and oil. While this process produces a single-grooved wheel more cheaply than by turning, it is doubtful whether from a manufacturing standpoint there is any economy in the built-up wheel when we consider the degree of accuracy now obtained in moulding multiple-grooved wheels, and the consequent reduction of labor in turning them.

Another special form of rope-driving pulley is that made by John Musgrave & Sons, Bolton, England. This pulley (Fig. 64), it will be noticed, is extremely light, but is sufficiently strong for the requirements. This lightness is attained by the use of steel arms turned tapering, and firmly secured to the rim and the hub, which latter is split in three segments and ringed with steel. By the use of this form of pulley the shafts are relieved to a considerable extent of the weight and consequent friction entailed by the ordinary pulley, and there are no excessive shrinkage strains to contend with, as is usually the case with the

FIG. 64.—ROPE PULLEY WITH TURNED STEEL ARMS.

common form of single-casting pulley. In pulleys of this form, where wrought-iron and steel rods are used for arms, the ends of the rods should be dipped in acid and tinned before setting in the mould. In addition to this the rods are frequently headed up and grooved used in the ends.

The bottom of the groove in cast-iron pulleys is sometimes filled with wooden blocks dovetailed into a channel cast to receive them, in which case the rope runs on the bottom and the shape of the groove approximates that shown in Fig. 65; after being fitted and secured the groove is trued up and turned out to the desired shape. Gutta-percha, rubber, leather, tarred hemp, and other materials have been used for the same purpose; but in the best modern practice a smooth cast-iron surface is preferred to any other, and we find these groove linings confined chiefly to pulleys used in wire-rope transmissions and hoisting-machinery.

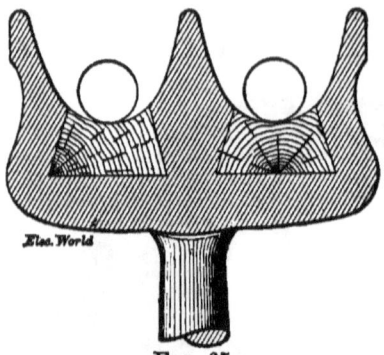

FIG. 65.
WOOD-FILLED PULLEY RIM.

Of late years an all-wood rim with V or U grooves has been used to some extent, as it makes a cheap pulley, and is very satisfactory for light work when a semicircular groove is adopted. Wooden rims, however, should not be used for heavy work, and the slip of the rope should be reduced to a minimum; otherwise the grooves and ropes will be rapidly cut out and the whole system will be very unsatisfactory.

The proportions of rope-pulley rims depend somewhat upon the shape of groove adopted. Some manufacturers have a standard groove, with straight sides, which may be used for several different-sized ropes, as shown in Fig. 66. For light work such a pulley is very satisfactory, and has many advantages in regard to constructive features not possessed by any other form. The groove shown in Fig. 67* is very commonly used in England, but in this country

---

* Low and Bevis, "Machine Design," p. 156.

manufacturers of large rope-pulleys usually prefer a form in which the abrupt change in profile (as at *a*) does not occur. Of these the more common form is a modification

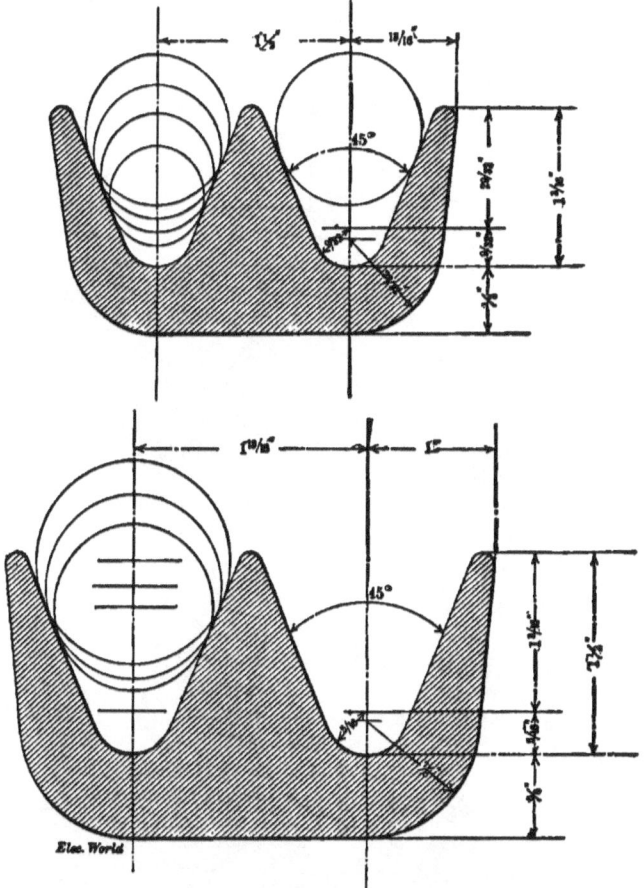

FIG. 66.—RIM SECTIONS—STRAIGHT GROOVE.

of the English section, in which the straight sides forming the angular groove are connected to the rib between the grooves by curves as in Fig. 68.

Another form which has much in its favor, especially for the independent rope system, is the circular arc groove, in which the sides are formed by arcs of circles.

Working proportions for this groove are given in Fig. 69, in which the unit d equals the diameter of rope. It

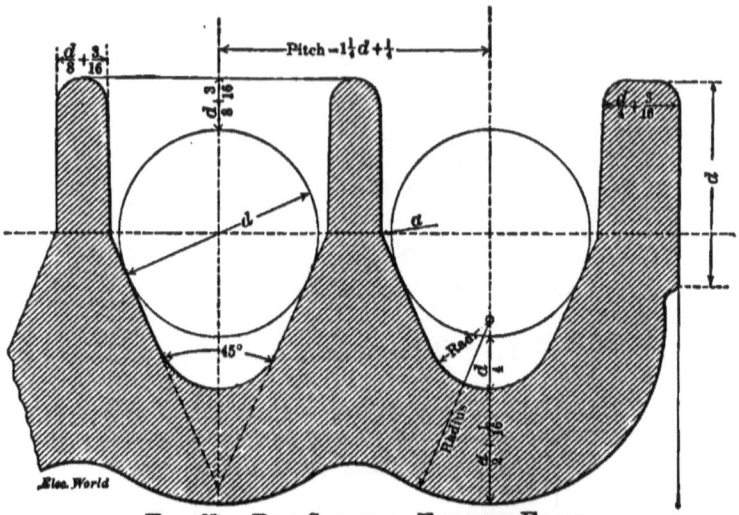

FIG. 67.—RIM SECTION—ENGLISH FORM.

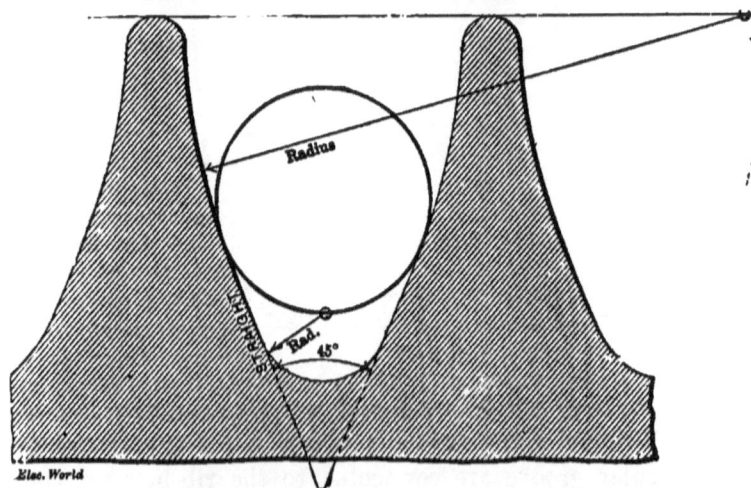

FIG. 68.—RIM SECTION—IMPROVED ENGLISH FORM.

will be noticed that the centre for the curve is located at the intersection of a line drawn through the centre of the rope at an angle of 22½° with the horizontal and a line drawn through the tops of the dividing ribs; the angle of the groove embraced by the rope is thus dependent upon the position of the latter in the groove: in its normal position the angle is one of 45°. The Walker groove, as now used in rope-pulleys made by Fraser & Chalmers of Chicago and the Walker Mfg. Co. of Cleveland, has its sides formed with circular arcs similar to Fig. 69, but the angle of the

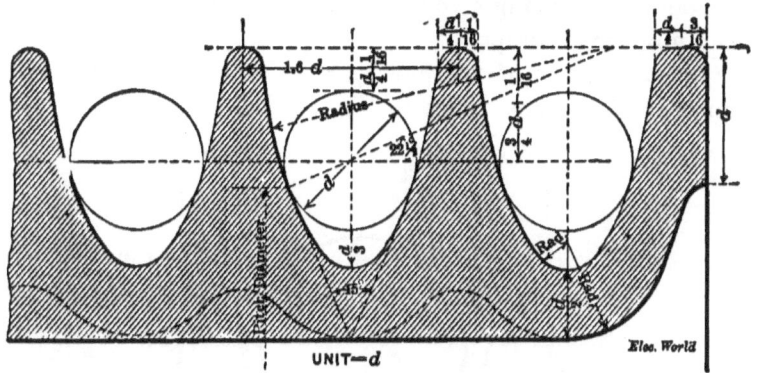

UNIT=d

ANGLE OF GROOVE 45° WHEN ROPE IS IN ITS NORMAL POSITION

FIG. 69.—RIM SECTION—GROOVES WITH CIRCULAR ARCS.

groove is more acute; at the point of tangency when the rope simply rests in the groove the angle measures 33°.

Rope-pulleys for ordinary transmissions are usually flat on the inside of the rim, or slightly tapered as shown by the full lines in Fig. 69. In large wheels, however, the rim is frequently swept up, and where the weight with a flat rim would be greater than required for strength or steady running, the inside is hollowed out as shown by dotted line. This gives a more nearly uniform section and makes a stronger and lighter wheel, but the expense of construction is greater. Guide-pulleys or idlers should be made with a semicircular groove, so that the rope runs upon the bottom instead of being wedged between the sides, as in grip-pulleys. Some engineers maintain that the wedge groove should be used in all cases, and many plants are in operation having V-shaped grooves in the idlers; but we believe the most satisfactory results will be obtained by using a groove in which the rope runs on the bottom. Of these there are two general forms—the one in which the rope has considerable play in the groove, and the other of such form that the rope is embraced by a portion of the surface of the groove: the first is used more particularly for single ropes, while the latter is adapted to any number of wraps.

A modification of the first form is frequently used for a number of ropes as shown in Fig. 70.

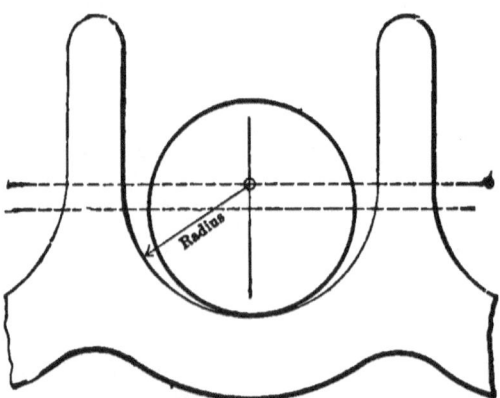

Fig. 70.—Form of Groove for Guide Pulleys.

The proportions shown in Fig. 71 will be found very satisfactory for single or multiple groove guide-pulleys, in which the pitch is equal to that used with the grip-pulleys represented in Fig. 69; the general design conforms closely to the latter, but the pulley is somewhat lighter.

For shaft-pulleys and the smaller sizes of fly-wheels not exceeding about nine feet in diameter the casting is usually made in one piece, unless a "split" pulley is required. In order to relieve the wheel in cooling the arms were formerly given a curved or S shape, as this form yields more readily and is supposed to conform to the unequal contraction of the rim and hub of the pulley in cooling. With properly proportioned wheels, however, and with due care in the foundry, straight-armed pulleys may be cast as strong as those of curved outline, and as the former are lighter, neater, and cheaper than the latter they are now almost universally employed. In the larger single-casting pulleys the hubs are frequently split in order to favor the arms in cooling:

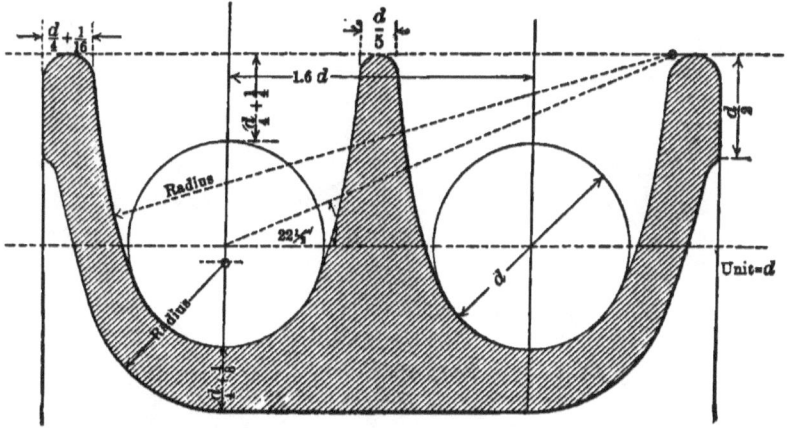

FIG. 71.—FORM OF GROOVE FOR GUIDE PULLEYS.

when split by a diametral plane the two portions at the hub are generally secured by bolts and nuts as in Fig. 72, or less frequently by pins and cotters; when split in three, wrought-iron or steel rings are shrunk onto the ends of the hub, which is turned to receive them, as represented in Fig. 73. Occasionally rings are shrunk on and bolts used as well, but this precaution is not common practice.

There is no general rule by which the number of arms may be determined. For small rope-pulleys the number is usually six, while for the larger sizes six or eight arms are used for pulleys cast in one or two parts; but these, in some cases, have two sets of arms as shown in Fig. 74, which represents a 78-inch pulley made by the Robert Poole & Son Co. for the Providence Cable Tramway Co. The usual

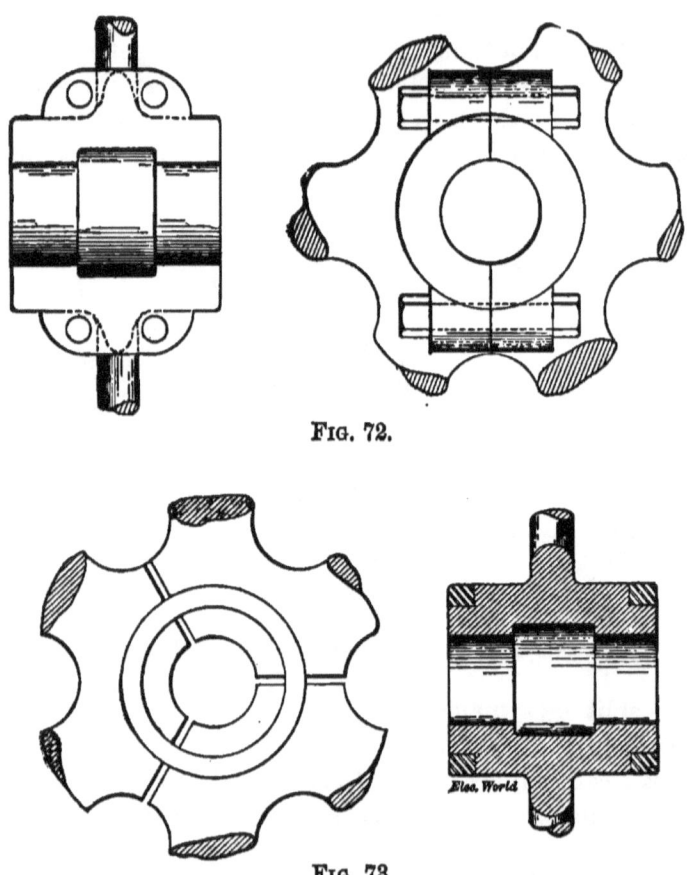

FIG. 72.

FIG. 73.

run of wheels from 9 feet to 15 feet in diameter are cast in halves, six or eight arms being employed.

Occasionally, however, narrow rope fly-wheels, of diameters up to 18 feet, are made in two pieces only, with eight arms.

ROPE-DRIVING. 193

Moderate-sized wheels are frequently cast in one, but arranged so as to be readily split in two afterwards.

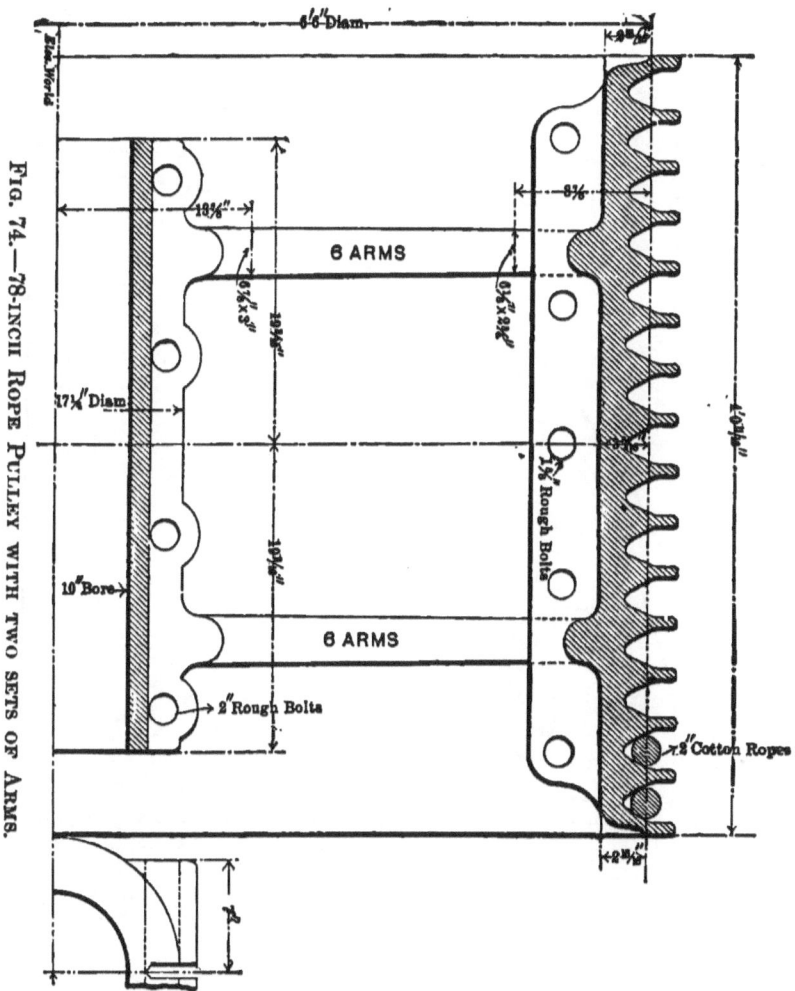

FIG. 74.—78-INCH ROPE PULLEY WITH TWO SETS OF ARMS.

Pulleys cast in halves having light rims should have double arms along the line of separation, as shown in Fig. 75, which represents a form of split pulley made by Fair-

ROPE-DRIVING.

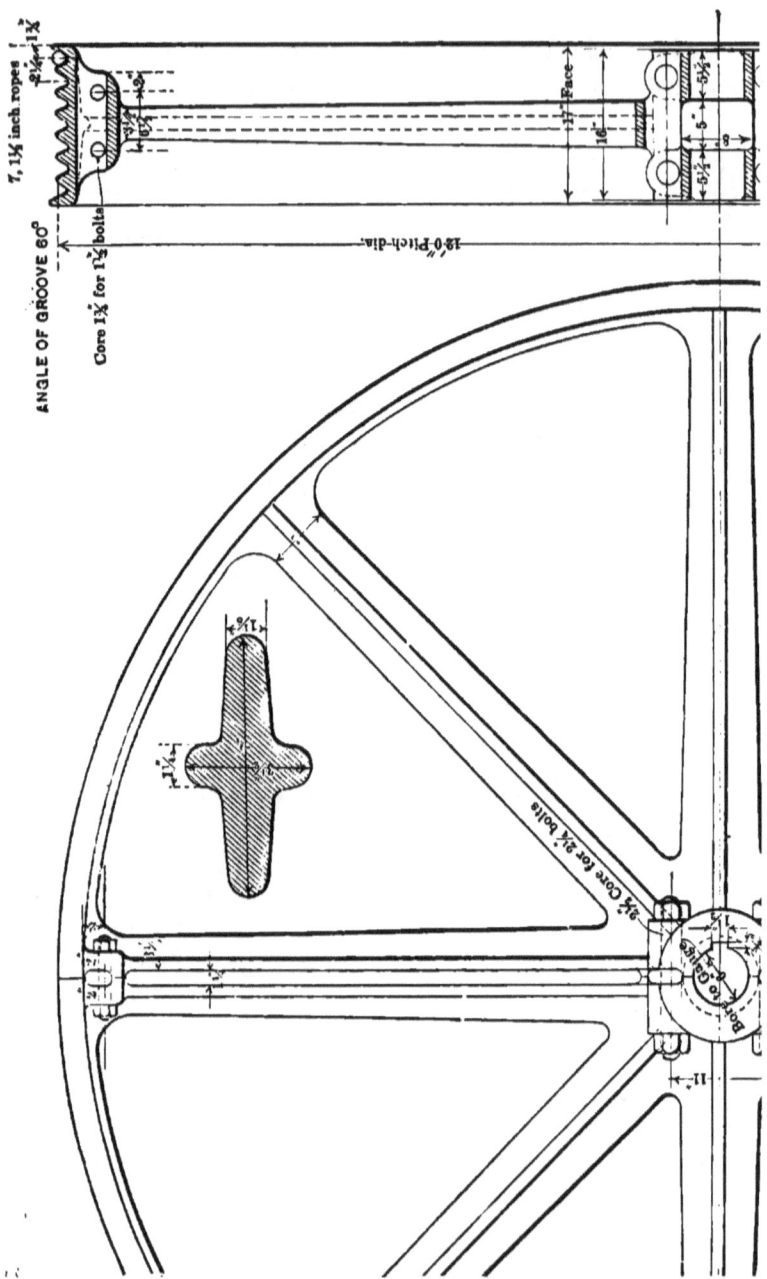

banks, Morse & Co.; this prevents undue bending action, and is otherwise the better able to resist the effect of centrifugal force due to the added weight at the point of connection.

For diameters ranging from 15 feet to 20 feet the usual practice is to make the pulley in six or eight segments with as many arms, although some makers prefer ten segments; from 20 to 26 feet ten segments are usually adopted. For wheels from about 26 to 32 feet twelve segments and as many arms are generally used, while larger-sized wheels are built up of fourteen, sixteen, and even more segments—sometimes as many as twenty-four being employed.*

When rope-pulleys are made in halves or are built up it is important that the connecting bolts and flanges in the rim should be strong enough to resist the maximum stresses that may occur in the joint.

Ordinarily in rope-pulleys the net section of bolt area is about 12 per cent of the section at the joint, but this ranges from 25 per cent to about 6 per cent in different cases. Various shop rules are used by designers for obtaining the bolt section, many of which give the bolt area directly in terms of the cross-section of the rim. While such formulæ may give satisfactory results for an assumed average rim speed, it is better to design the joint in large wheels from a consideration of the particular conditions in each case.

In very slow-moving heavy wheels the principal stress may be that due to the weight of the wheel itself, but in rope-pulleys and fly-wheels the rim has usually a high velocity, and the strain due to centrifugal force in the rim is the principal factor in determining the bolt section; it is obvious, however, that in high-speed heavy wheels both influences should be taken into consideration.

The tension in the rim produced by centrifugal force is

---

* *American Machinist*, Feb. 16, 1893.

equal to one half the centrifugal force due to the weight and velocity of the rim multiplied by the ratio of diameter to semi-circumference, or equal to

$$\frac{F_0}{2} \times \frac{2r}{\pi r} = \frac{F_0}{\pi}.$$

As this force is resisted at each end of a diameter, the strain T_h, or hoop-tension, acting at either end will be one half the above; hence

$$T_h = \frac{1}{2}\frac{F_0}{\pi},$$

in which F_0 has the usual value $0.00034\, WRN^2$,
 where $W =$ entire weight of rim;
 $R =$ effective radius of rim in feet;
 $N =$ revolutions of wheel per minute.

The tension at each end of a diameter due to the weight of the rim is evidently equal to $\frac{1}{2}\left(\frac{W}{2}\right)$; therefore the stress in the rim due to F_0 and W will be

$$S = \frac{1}{2}\left(\frac{F_0}{\pi}\right) + \frac{1}{2}\left(\frac{W}{2}\right) = \frac{1}{2}\left(\frac{F_0}{\pi} + \frac{W}{2}\right).$$

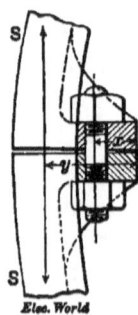

Fig 76.

If the bolts could be placed at the point of application of this force, their effective section A would be $\frac{S}{f}$, where f is the allowable stress in the bolts.

In the actual construction the centre of bolts must be placed at some distance from the point of application, in which case the bolts are subjected to an additional strain due to the bending moment at the joint, as shown in Fig. 76. If this bending moment is taken up by the rim, the bolt section will be

$\dfrac{S}{f}$ as before; but if the rim and bolt flanges are rigid and resist any deformation, the total strain comes upon the bolts. Under these conditions the net section A may be obtained by multiplying $\dfrac{S}{f}$ by the ratio of the leverage y of the force S to the leverage x of the resistance in the bolt; hence if we assume that $x = \frac{1}{2}y$, a common value, we shall obtain $A = 2\dfrac{S}{f}$.

By using studs with nuts at each end the bolt may be placed nearer the rim, in which case, for the same depth of flange, the ratio $\dfrac{y}{x}$ may be made less than 2. In order to obtain ample strength at the rim, the bolt section should be determined on the supposition that the full bending moment will be thrown upon the bolts. In this case the bolt-flanges or lugs must be able to resist any bending due to the force S. The bolt-flanges will be as strong as the rim if we make the respective moments of resistance equal to each other; hence

if b = face of wheel;
   b' = breadth of bolt-flange minus the width of bolt-holes;
   t = mean thickness of rim;
   t' = thickness of bolt-flange or lug,
then

$$\tfrac{1}{6}bt^2 f = \tfrac{1}{6}b't'^2 f \quad \text{or} \quad t' = \sqrt{\dfrac{bt^2}{b'}}.$$

This thickness t' may be reduced somewhat if strengthening ribs are carried from the rim to the lower edge of the bolt-flange, as shown in Fig. 76. In the above, t was taken as the mean thickness of rim reduced to a rectangle, and no account has been taken of the strengthening ribs and flanges. If the leverage of the bolts is assumed as one half, we may obtain the total net section of the bolts as follows:

$$A = 2\frac{S}{f} = \frac{1}{f}\left(\frac{F_0}{\pi} + \frac{W}{2}\right).$$

Since $F_0 = 0.00034\,WRN^2$,

$$A = \frac{1}{f}\left[0.000108\,WRN^2 + \frac{W}{2}\right].$$

In the above formula f may be taken equal to 6000 pounds, for bolts up to $1\frac{1}{4}$ or $1\frac{1}{2}$ inches diameter, but for larger sizes 8000 to 9000 pounds may be used.

If $t =$ the mean thickness and $b =$ breadth of rim in inches, the weight W may be determined from

$$W = 2\pi R \times 12tb \times 0.26$$
$$= 19.6\,Rtb.$$

If the rim is properly bolted, the principal strains to which the hub-bolts are subjected are those due to the weight of hub and arms and the tension produced by keying to the shaft.

As the weight of hub and arms in large pulleys is usually much less than the weight of rim, it will be seen that the strain on the hub-bolts due to the weight and centrifugal force of these parts will be less than that on the rim-bolts.

In practice some makers design the hub-joint so that the net section of all the bolts is equal to the bolt section in one edge of the rim-joint; this, however, is not usual practice. An inspection of a great many pulleys made by various manufacturers shows that the hub-bolt section is often twice as great as the rim-bolt section; but in many of these cases the rim-bolting is very weak.

It is safe to make the bolt area the same in each case; but if the rim-joint is made according to the method

ROPE-DRIVING.

indicated, the bolts in the hub will usually be sufficiently strong if their total effective section is equal to that in one rim joint: for light pulleys, however, this should be increased.*

The section of arm for those pulleys in which the arms and hub are cast together is usually elliptical or segmental, as shown in Fig. 77.

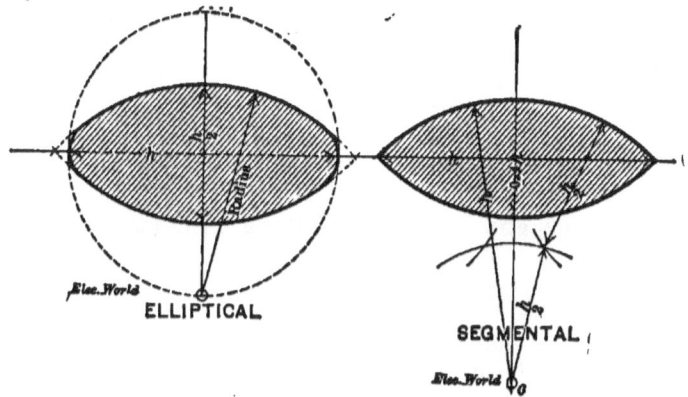

FIG. 77.—SECTIONS OF ARMS.

The elliptical form as here given is so proportioned that its minor axis is one half its major; in the segmental form the minor axis is four tenths the major.

The cross-section of the arm may be determined by considering it as a beam fixed at one end and loaded at the other with the force P due to the pull of the ropes. The bending moment on each arm will then be

$$M_b = \frac{Pr}{N_a},$$

assuming that the load is divided equally among the number of arms, N_a.

---

* For complete analysis of strains in fly-wheel rims, see Prof. Unwin's discussion in his "Machine Design," vol. II. Also. Mr. Stanwood's paper in vol. XIV. Trans. A. S. M. E., with Prof. Lanza's discussion.

Since the bending moment is equal to the modulus of the section, Z, multiplied by the stress, f, in the material, we have

$$M_b = fZ;$$

For an elliptical section whose major axis is h and minor axis $0.4h$,

$$Z = \frac{\pi}{32}h^2 \times .4h;$$

hence

$$\frac{Pr}{N_a} = f \times \frac{\pi}{32} \times .4h^3. \quad \ldots \ldots (18)$$

If we let T_1, the maximum tension in the rope, equal $200d^2$ pounds, and $\frac{T_1}{T_2} = 2$, we shall have the load acting on each arm

$$P = 100d^2n,$$

where n is the number and d the diameter of ropes. Therefore by substituting this value of P in equation (18) there is obtained $h = \sqrt[3]{\dfrac{100d^2nr}{0.04fN_a}}.$

Assuming that $f = 2500$ pounds per square inch for cast iron and that N_a equals 6, we have

$$h = 0.44\sqrt[3]{d^2nD},$$

where D = pitch diameter of pulley and h is the major axis of arm produced to centre of wheel, both in inches. As the smaller pulleys require a larger margin of strength, on account of their greater liability to breakage in handling and casting, the above formula should be modified by introducing a constant; if we let this constant equal $\frac{1}{2}$ inch, a suitable value for the width of arm will be obtained from

$$h = 0.44\sqrt[3]{d^2nD} + 0.5''. \quad \ldots \ldots (19)$$

As the centrifugal force set up in the rim of the pulley causes an additional stress in the metal, the value of f should be chosen with reference to the speed at which the pulley is to run. The higher speed not only induces a greater stress in the material, but the liability of failure due to vibration or shocks is greatly increased. The stress in the rim due to centrifugal force may be obtained by considering the pulley as a cylinder subjected to a force $p(= F_o)$ per square inch. The thickness of a thin cylinder to resist rupture may be obtained from

$$t = \frac{pD''}{2f'} = \frac{pr}{f'},$$

in which $t =$ thickness in inches;
$p =$ pressure per square inch;
$r =$ radius of cylinder in inches;
$f' =$ allowable stress in pounds.

In this case

$$p = F_o = \frac{Wv^2}{gR};$$

$W =$ weight in pounds $= 0.261$ pounds per cubic inch;
$v =$ velocity of rim in feet per second;
$g = 32.16$;
$R =$ radius of rim in feet $= \frac{r}{12}$;

hence

$$t = \frac{F_o r}{f'} = \frac{Wv^2}{gR} \times \frac{r}{f'} = \frac{Wv^2 r}{gf'\frac{r}{12}}$$

and

$$f' = \frac{12Wv^2}{gt}.$$

If F_o act on one square inch of pulley whose thickness is unity, we shall have

$$f' = \frac{0.261 \times 12v^2}{32.16} = 0.097v^2$$

$$= \frac{v^2}{10}, \text{very nearly};$$

this is the stress in rim per square inch of section due to centrifugal force alone.

When v in feet per second =	50	60	70	80	90	100	150	200
" V " " " minute =	3000	3600	4200	4800	5400	6000	9000	12000
Stress f' in pounds due to centrifugal tension =	250	360	490	640	810	1000	2250	4000

The stresses due to the pull of the ropes, and those due to contraction in cooling, are additional to those here given, hence f should be taken sufficiently low to allow for the various stresses which may be set up in the pulley. If we assume that the working stress should not ordinarily exceed 3500 pounds per square inch for cast-iron pulley arms, and that it should be less for high speeds where the dynamic effect of shock and vibration is greater, a suitable value for cast iron may be obtained from the empirical formula

$$f = \frac{50000}{5 + \sqrt[3]{V}}.$$

From this formula the annexed values of f have been calculated; it will be noticed that the value of f used in formula (19), namely, 2500 pounds per square inch, corresponds to a velocity of rim equal to about 3500 feet per minute.

Velocity of rope in feet per minute $V =$	1200	1800	2400	3600	4800	6000
Allowable stress $f =$	3200	2900	2700	2450	2300	2150

ROPE-DRIVING. 203

The arms should taper toward the rim $\frac{1}{16}$ inch per inch of length, that is, $\frac{1}{32}$ on each side, but in no case should the width of arm at the rim be less than two thirds its width at the centre of shaft.

Very wide pulleys in which the proportions for single arms would be inconveniently large may be made with two or three sets of arms; in such cases they may be considered as two or three separate pulleys combined in one, except that the proportions of the arms should be 0.8 to 0.7 times that of single-arm pulleys.*

In designing the hubs of wheels practice varies considerably. Some authorities give the thickness of metal around the eye in terms of the pulley diameter only, others take into account the diameter of pulley and also the diameter of shaft or the breadth of face, while the length is variously given in terms of the diameter of shaft or the face of pulley, or both.

For rope-pulleys if the thickness of metal in the hub is made proportional to the diameter of pulley and also to the diameter of shaft, and if the length of hub is made to vary with the number and size of ropes and the diameter of shaft, we believe the requirements of strength and good proportions will be best attained.

If D = diameter of pulley,
 $d_s =$ " " shaft,
 $d_h =$ " " hub,
 $d =$ " " rope,
 $n =$ number of ropes,
 $L_h =$ length of hub,

then the diameter and length of hub may be obtained from the following formulæ, which have been deduced from the proportions of a large number of rope pulleys made by representative manufacturers:

$$d_h = 0.025 D + d_s + 2\tfrac{1}{4}''$$

---

* Reuleaux, "Constructeur."

and
$$L_h = 0.6dn + d_s + \tfrac{3}{4}''.$$

For loose or idle pulleys the diameter of hub may be made less than that given by the above formula, which allows for keying. In general the thickness of metal around the eye in an idle pulley may be taken as about two thirds as great as that in fixed pulleys of the same diameter and face. The length of hub in idlers should be sufficient to give a good bearing surface, and may vary from two to three times the diameter of shaft—depending somewhat upon the speed of rotation.

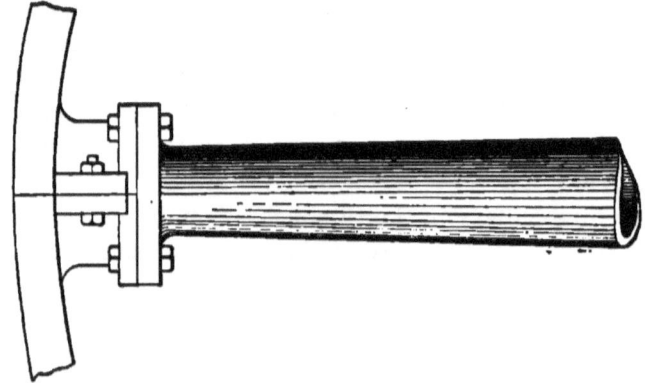

FIG. 78.—METHOD OF JOINING ARMS AND RIM.

In the construction of large rope-pulleys which are made in segments, the usual method is to bolt the rim section to the arms at the ends of the segment, as shown in Fig. 78.

When the rim segments are joined midway between the arms as in Fig. 79 the several connections are simplified to some extent, but with this method of connection there is a decided tendency for the joint to open under the influence of centrifugal force, which is increased somewhat by the weight of the connecting flanges and bolts. With the rim-joints at the junction with the arms, however, a rigid connection between adjacent arms is obtained, and the centrifugal effect of the rim tends to increase the tension in the arm without opening the joint.

Examples of both methods of connection are given in Figs. 80 to 92, which also represent some of the details of construction in various built-up rope-pulleys.

As will be noticed, these built-up wheels are made with a large central boss or hub, usually in one piece, but sometimes in two, provided with seats for the arms, or bored,

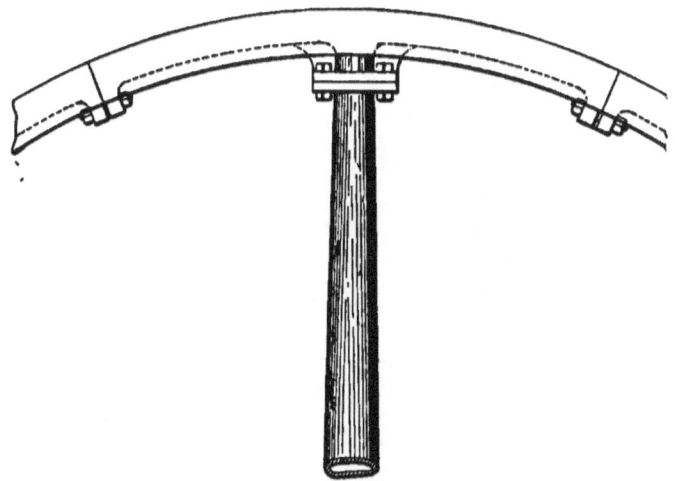

FIG. 79.—METHOD OF JOINING ARMS AND RIM.

either tapering or straight, to receive the ends of the arms, which are turned to fit the holes.

The arms are usually of round or elliptical section, cast hollow; but other forms are common. Thus in Figs. 85 and 90 the arm is of cruciform section, while in Fig. 81 a modification of the H section is used.

In Fig. 80 we have an interesting form of wheel patented some years ago by Mr. James Barbour, of Combe, Barbour & Combe, Belfast.* The rim of the wheel is made in seg-

---

* *Engineering*, Sept. 7, 1888.

ments, and is attached to the boss or hub by means of the bolts or tie-rods d, extending from the rim to the boss as shown, and passing through the arms, which are made hollow for the purpose. The rim and hub are recessed to receive projecting pieces $a'a'$ on the ends of the arms. The

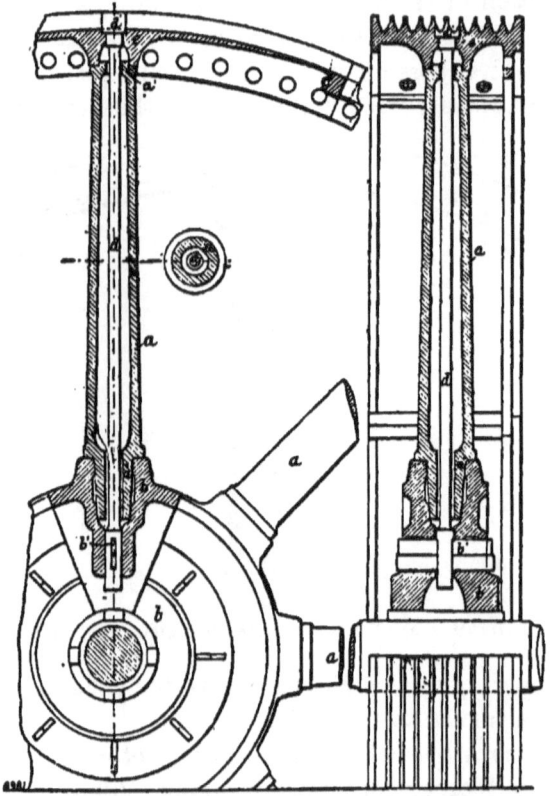

FIG. 80.—BUILT-UP WHEEL. (Barbour.)

head of the tie-rod d' can be made to form part of the periphery of the wheel. The tie-rods have slots formed in their inner ends to receive keys b', which keys are passed into the hub through holes parallel to the axis of the shaft, and in this way the rim of the wheel is secured to the hub.

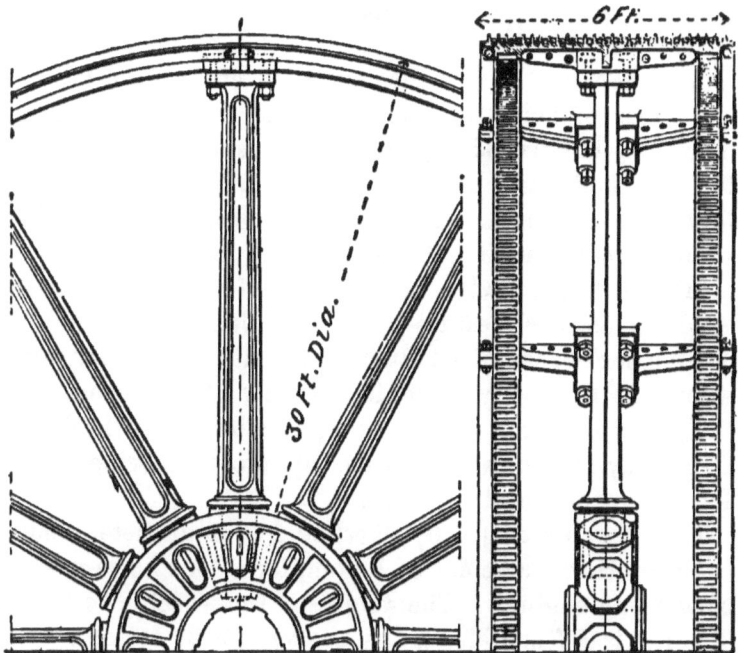

Fig. 81.—Thirty-foot Rope-wheel. (Hick.)

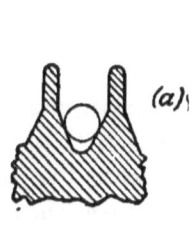

Fig. 81 (a).
Section of Groove.

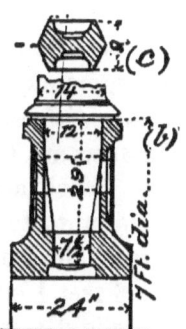

Fig. 81 (b) and (c).
Section at Boss.

The object of the arrangement is to obviate the use of a number of bolts for connecting the rim to the arms; moreover, the tie-rods withstand the centrifugal force of the rim, and the wheel can be driven with safety at a high rate of speed. The usual projections from the principal surfaces are also eliminated, and the currents of air generated by the high velocity of the rim are lessened. This latter feature is one the value of which is becoming more recognized lately, especially in rope-driving, when a high velocity of rim is especially advantageous.

To lessen the fan action set up in a large fly-wheel, we frequently find the arms boarded up, provision being made for this in the wheel.

Fig. 81 represents a fly-wheel 30 feet in diameter made by Hick, Hargreaves & Co., Bolton, for a 1000 horse-power condensing-engine.* The weight of the wheel is 54 tons. The rim is 6 feet wide, and is grooved for 27 cotton ropes 5 inches in circumference, or nearly $1\frac{5}{8}$ inches in diameter, at $2\frac{1}{4}$ inches pitch. The grooves are shown in detail at (a).

The rim is constructed of twelve segments with twelve arms. The segments are planed at the joints, and are bolted together with eight bolts and nuts, of which the two next the arms are $1\frac{3}{4}$ inches in diameter and the others are $1\frac{1}{2}$ inches. Each arm is secured with four 2-inch T-head bolts and nuts to the rim segments, two to each of two segments. The boss, nave, or centre, shown in section at (b), is 6 feet $\frac{1}{4}$ inch in diameter, or 7 feet across the platforms, or slightly raised plane surfaces, on which the arms take their bearings. It is 18 inches wide at the rim and 2 feet wide at the bearing on the shaft. Twelve sockets are bored out to receive the ends of the arms. The arms are formed approximately of H section, as at (c), and

---

* D. K. Clarke, "Steam-Engine," vol. III.

measure 14 inches by 9 inches at the centre and 9¼ inches by 6¾ inches at the rim. They are turned conically to fit

Fig. 82.—Twenty-four-foot Rope-wheel. (Walker.)

the holes in the centre, tapering from 12 inches in diameter at the outer ends of the sockets to 7½ inches at the inner ends in a total length of 2 feet 5 inches. The arm is keyed into the centre with two cotters, each 24 inches long, 1 inch thick, tapering in width from $3\frac{3}{4}$ inches to $3\frac{1}{4}$ inches, or $\frac{1}{2}$ inch in 24 inches. They are driven in one from each side of the nave reversely, and make up a united width of $6\frac{3}{4}$ inches. To join the rim the arm is expanded into a flat flange 21 inches by 18 inches and 4 inches thick, through which the four bolts already named are passed. The opening at the centre is 25 inches in diameter, or 2 inches larger than the shaft. The wheel is fixed on the shaft with six keys 5 inches wide and $1\frac{5}{8}$ inches thick, bearing on six flat seats formed on the shaft, with a taper of $\frac{1}{8}$ inch to the foot.

Fig. 82 represents a somewhat similar method of connection employed by the Walker Manufacturing Company, Cleveland. In this pulley the arms are of round or pipe section, and present a much neater appearance and better proportions than obtain in the previous pulley.

This wheel was made for the Baltimore City Passenger Railway Station, and is 24 feet in diameter, 30 inches face, grooved for ten 2-inch ropes at $2\frac{7}{8}$ inches pitch. The details are shown in Figs. 83 and 84.

The centre is a single casting 5 feet 6 inches in diameter, bored to receive the arms, of which there are ten. Each arm is $10\frac{1}{2}$ inches in diameter near the hub and 8 inches in diameter near the rim, where it is expanded into a flange and bolted to the rim segments. The bolts connecting the arms to the segments and the segments to each other are all turned $1\frac{15}{16}$ inches in diameter, and fit in drilled and reamed holes.

The shoulder on the arms at the hub is 15 inches in diameter, and is faced to fit the machined seats on the boss. By this means perfect alignment is obtained, and no motion of the parts is possible unless the key shears.

Fig. 85 represents a rope-pulley designed by Mr. F. Van

ROPE-DRIVING. 211

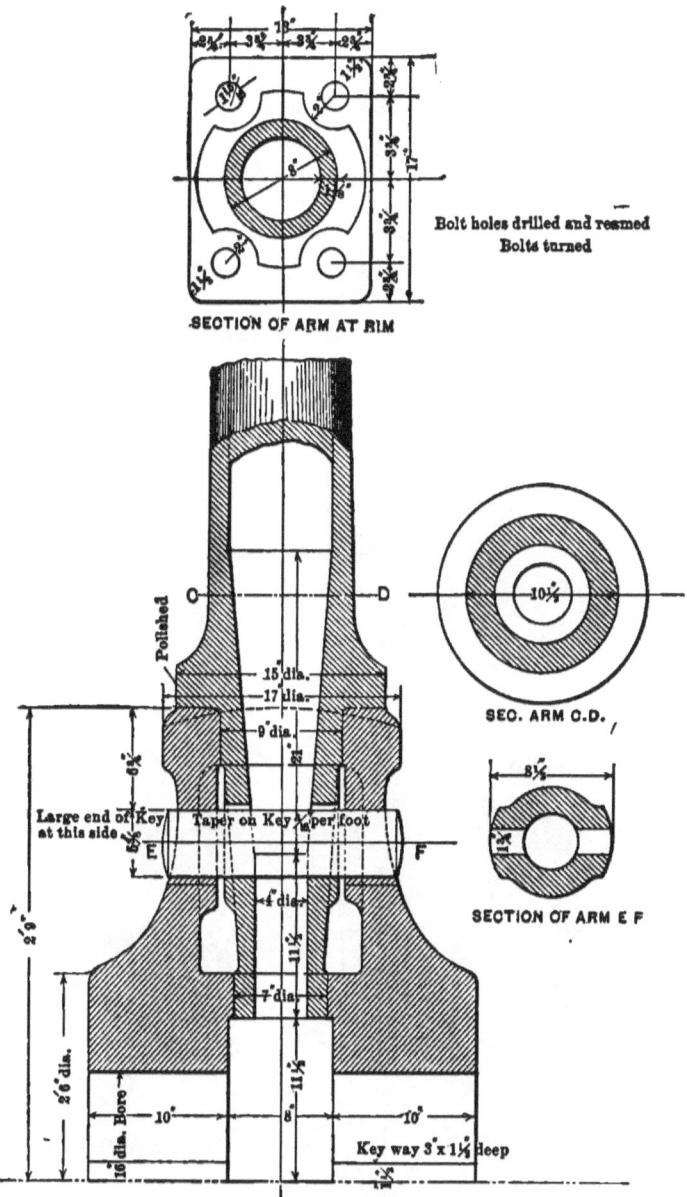

FIG. 83.—DETAILS OF PULLEY SHOWN IN FIG. 82. SECTION OF ARM.

212   ROPE-DRIVING.

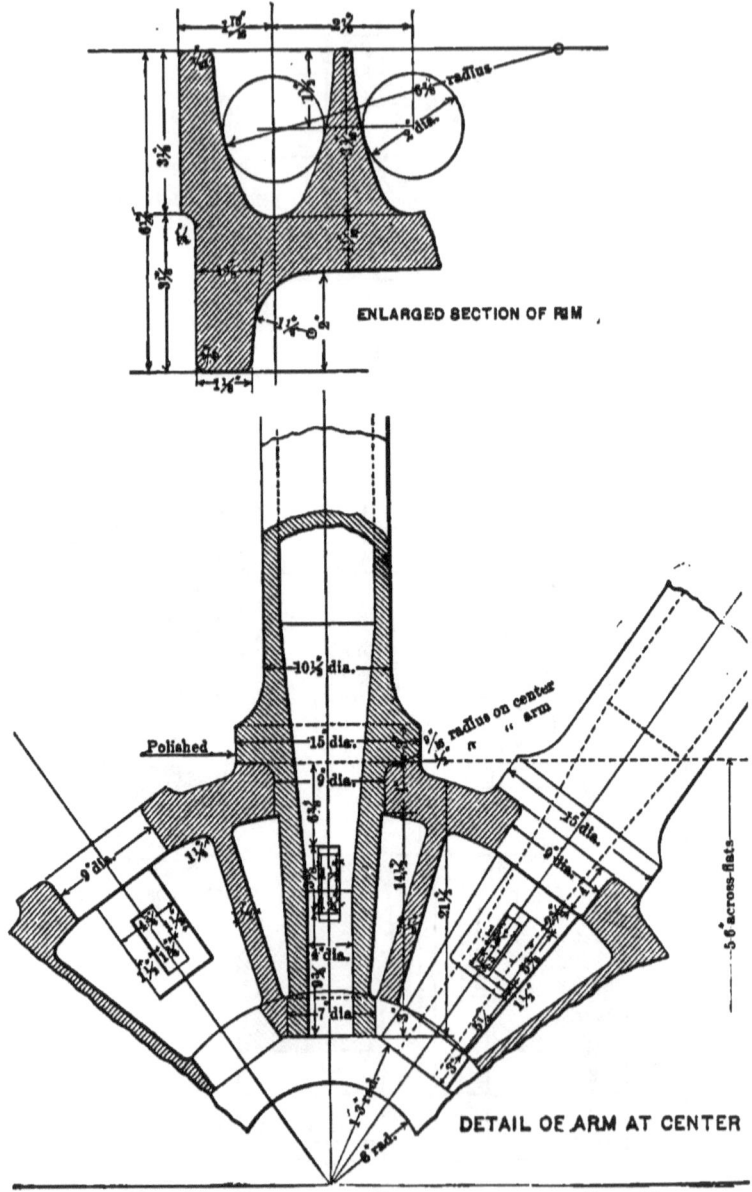

FIG. 84.—DETAILS OF PULLEY SHOWN IN FIG. 82.

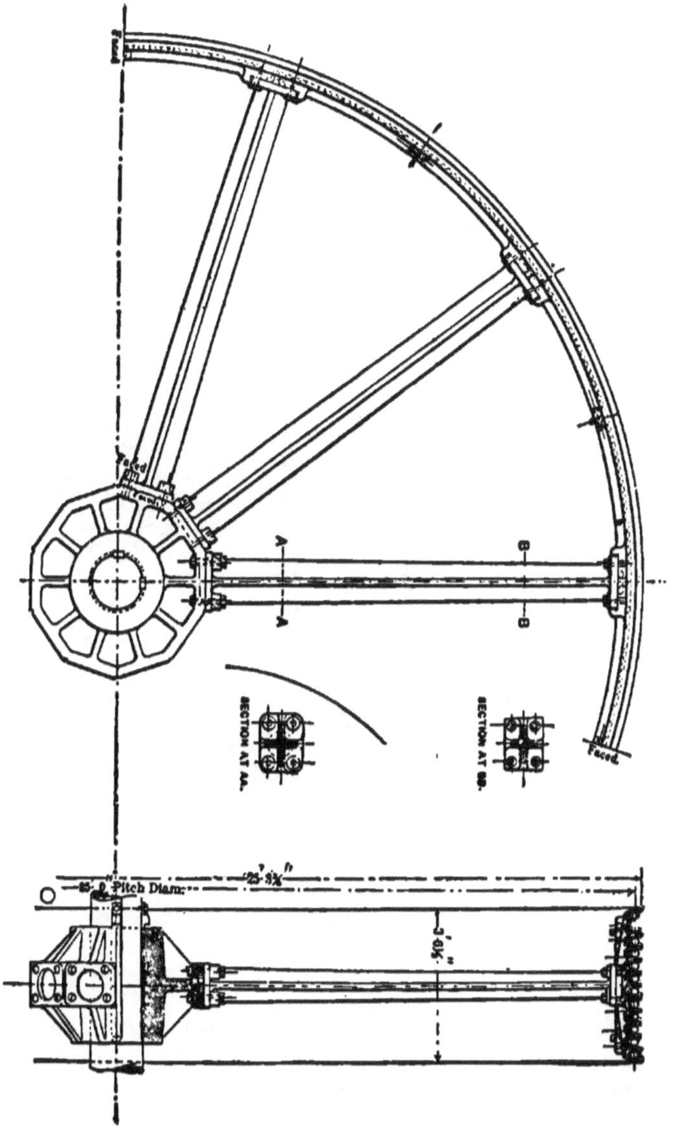

ROPE-DRIVING. 213

FIG. 85.—TWENTY-FIVE-FOOT ROPE-WHEEL. (Van Vleck.)

214   ROPE-DRIVING.

Vleck for the San Diego Cable-railroad Power-station.* It is 25 feet in diameter, 42½ inches face, grooved for twelve 2-inch cotton ropes at 3½ inches pitch. This drum is composed of ten separate segments and ten arms, bolted and keyed midway in each segment. The central hub is 4¼

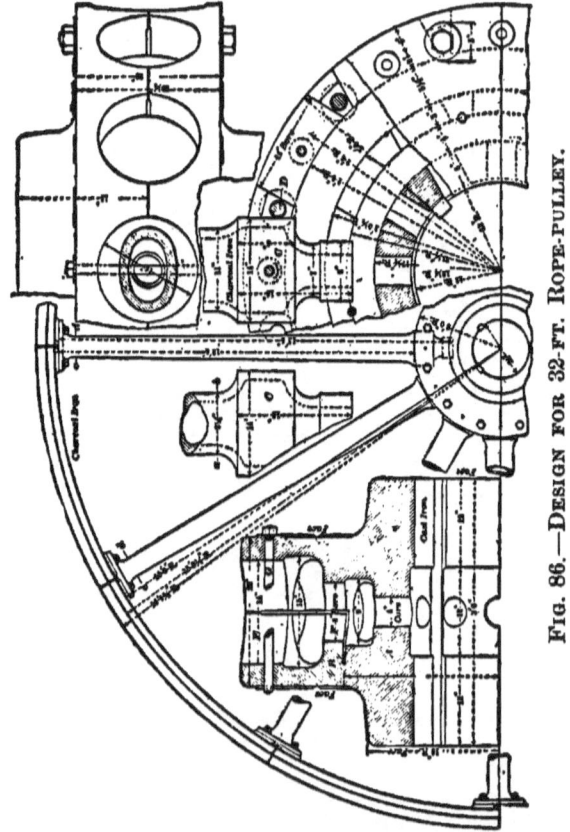

FIG. 86.—DESIGN FOR 32-FT. ROPE-PULLEY.

feet in diameter across flats, and is proportioned, as in fact the whole wheel is, with special regard to lightness.

The designs shown in Figs. 86 to 89 are for the large rope-pulleys used in the Fifty-first Street and Houston Street

---
* Trans. A. S. M. E., vol. XII. p. 77.

ROPE-DRIVING.

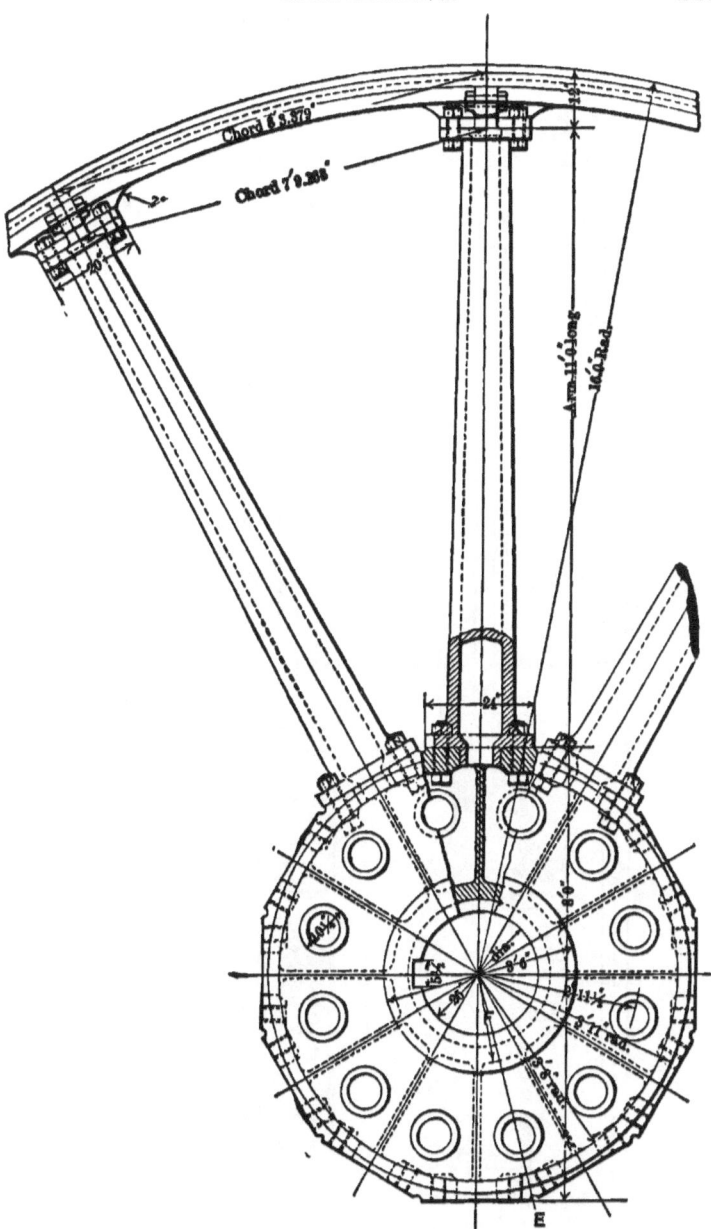

FIG. 87.—THIRTY-TWO-FOOT ROPE-WHEEL. (Walker Mfg. Co.)

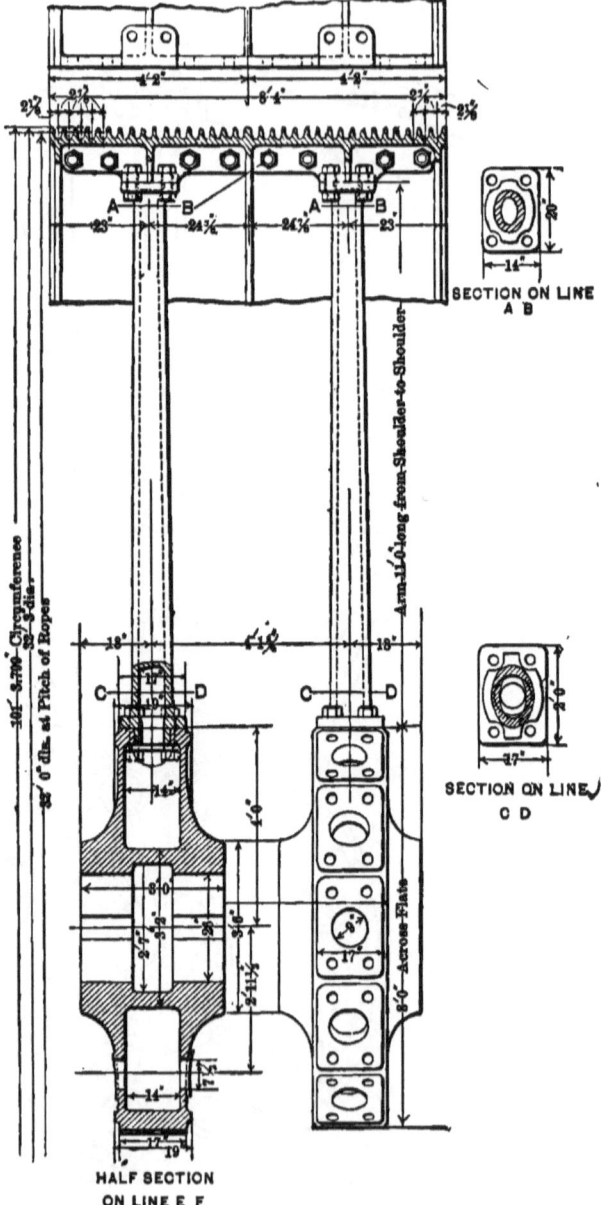

Fig. 88.—Thirty-two-foot Rope-wheel. (Walker Mfg. Co.)

stations of the Broadway Cable Road. The first wheel, Fig. 86, is 32 feet in diameter, 37½ inch face, and is grooved for 13 cotton (Lambeth) ropes 2 inches in diameter, 2¾ inches pitch. The speed of the ropes is very low—only 1877 feet per minute. Mr. M. W. Sewall* gives the following particulars regarding the details of this wheel, as at first designed. It will be seen by reference to the figure that the arms of the drum are secured to the centre by clamping two turned portions on the inner end of each arm, between heavy cast-iron disks, one of which is cast

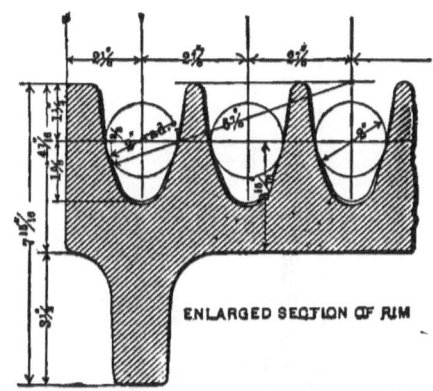

Fig. 89.—Section of Rim of Pulley Shown in Figs. 87 and 88.

as a flange on the hub and the other acts as a follower. These are bolted up a slight distance apart, and the holes for the arms are bored accurately to fit the turned portions of the same. The bolts are then loosened, the arms put in place, and the follower bolted hard up to the hub flange. A taper pin is then driven into a reamed hole through each arm and the parts of the centre, and held in place by a nut. The arms and the centre are then practi-

---

* For complete description of the cable-driving machinery see *American Machinist*, May 24 and 31, 1894.

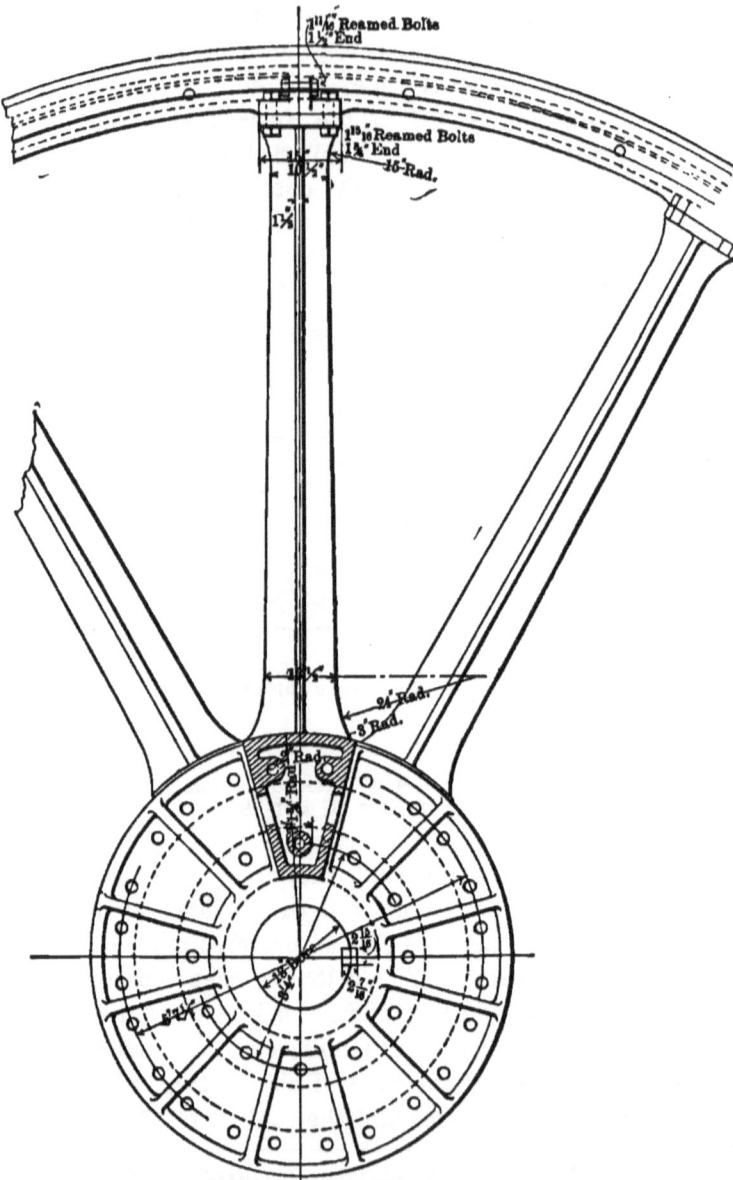

Fig. 90.—Heavy Rope-wheel, Twenty-seven feet Diameter.
(Wetherill & Co.)

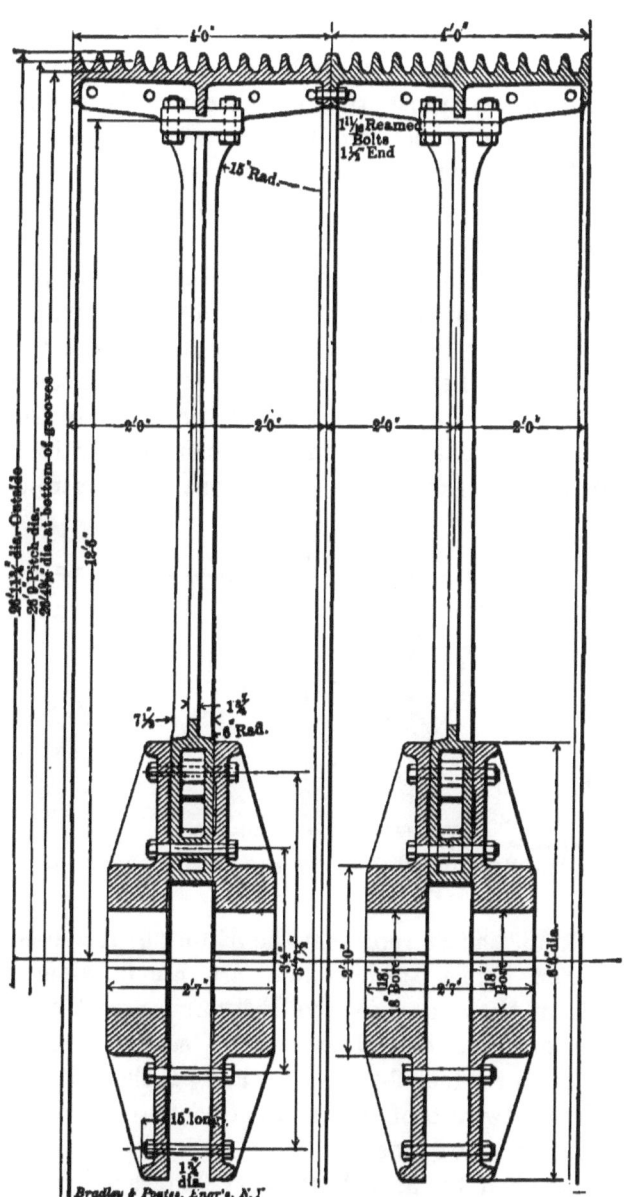

Fig. 91.—Section of Twenty-seven-foot Wheel.

cally one piece, and no working of the parts among themselves is possible. None of the bolts secure an individual arm, and no stress transmitted to the centre through the arms brings any of them into direct tension; several bolts might be broken without detriment to the structure. This drum would be improved, so far as convenience of manufacture is concerned, by a shoulder on the arm to determine its distance from the centre when assembling in the shop. The arms are attached to the rim in the usual manner, but are rather light in proportion to the rest of the design. These pulleys were redesigned to suit the contractor for the work, The Walker Manufacturing Co. of Cleveland, and were constructed in a manner similar to that shown in Fig. 87, except that single arms and centre were used.

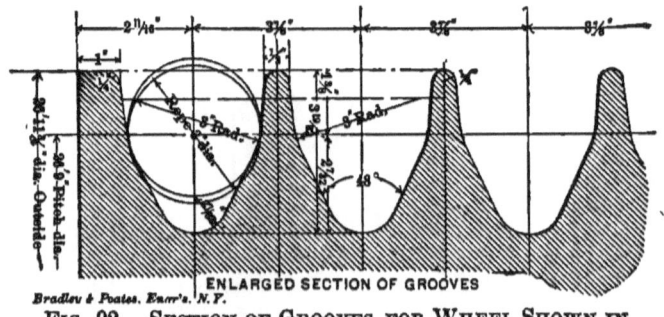

Fig. 92.—Section of Grooves for Wheel Shown in Figs. 90 and 91.

Figs. 87, 88, and 89 represent the 32-foot drums used in the Houston Street station, and are much wider and heavier than those used in the Fifty-first Street station. As will be noticed, these drums are 8 feet 4 inches face, grooved for thirty-four 2-inch ropes at 2¾ inches pitch, and arranged with two sets of arms and two centres. The weight of each pulley is 104 tons.

As constructed, the arms have heavy flanges bolted to the centre, into which a turned projection on the end of the arm is accurately fitted. In addition to this two of the four bolts securing each arm to the centre is turned to fit reamed holes in the boss; these bolts thus act as dowels and prevent any working of the parts. The bolts are calculated to resist the maximum tension in the arms, and are $2\frac{1}{16}$ and $2\frac{1}{2}$ inches in diameter; in the same way the arms are bolted to the rim, two of the four bolts fitting in reamed holes. The arms are of hollow elliptical section 15 by 10 inches at the hub, and $10 \times 7\frac{1}{2}$ inches at the rim. Each centre is 8 feet diameter across flats; they are bored 26 inches in diameter, and are 3 feet long in the bore.

Large rope-pulleys from 20 to 32 feet in diameter are, when extra wide, frequently made with two centres as well as two sets of arms, as just noted in Fig. 87. This is also seen in Figs. 90 to 92, which represent a rope-driving wheel 26 feet 9 inches pitch diameter, 8 feet face, grooved for twenty-four 3-inch ropes, at $3\frac{3}{4}$ inches pitch. This pulley was designed and built by Robt. Wetherill & Co. of Chester, Pa., and consists essentially of two separate and independent drums, flanged and bolted together at the rim. Each centre is made of two separate disks 6 feet 6 inches in diameter, bored and faced on the inside. The arms, of which there are twelve, are of cruciform section between the boss and the rim, where they are flanged and bolted to the rim segments in the usual manner. At the centre the arms are wedge-shaped, 8 inches thick, and are so proportioned that when accurately planed and fitted they form a complete circle.

These arm segments are then bolted between the two centre disks, and make a strong and compact hub.

INDEX.

Actual section of ropes, 92
Adhesion of ropes, 168
Advantage of high rotative speeds, 157
Advantages of rope-driving, 4, 5
American system of rope-driving, 24, 25
Angle embraced by rope, 117
Angle of groove, 164
Area of ropes, 92
Arms, number of, in pulleys, 190
Arms, shape of, 199
Arms, taper of, 203
Atlas Mills, 12
Atmospheric changes, 31
Automatic tension-carriage, 28
Axial rotation of ropes, 174

Barbour, James, built-up pulley, 205
Barrus, G. H., friction in mills, 7
Beeswax on ropes, 89
Belts, leather, 1, 34, 40
Blow-holes in pulleys, 183
Bolt area for pulleys, 195
Braided rope-joint, 23
Broadway Cable Road, built-up pulleys, 217
Brown, A. G., friction loss, 6
Brush Electric Light and Power Co., Niagara Falls, 56
Built-up pulleys, 183, 195, 205 to 221

Catenary curve, 8, 12
Cast grooves, 183
Centrifugal force, 45
Centrifugal force, influence of, 111, 119, 163
Combe & Barbour, early rope-drives, 2
Combe, Barbour & Combe, basis for calculating horse-power, 99

Cotton ropes, 77, 81
Cotton fibre, 78
Cotton wax, 78
Continuous-wind system, 25
Cone-pulleys for ropes, 35
Coil-friction, 44, 48
Corliss engine and use of jack-shaft, 41
Coefficient of friction, 112
Cost of ropes, 101, 122
Coulter, Dr. S. M., tests on manilla fibre, 82
Coupling, cut-off, 27
Coupling for braided rope, 23
Creep of ropes, 174
Cross-section of rope, 92
Cut-off coupling, 27

Deflection of rope, 131, 170
Degree of twist in ropes, 86
Details of rope-pulleys, 205 to 221
Diameter of bolts for pulleys, 195
Diameter of pulleys, 103, 161, 177
Differential driving, 36, 170
Differential pulley, Walker's, 173
Double arms in pulleys, 193
Double ropes recommended, 108
Draw-rods, 71
Durie, James, on rope-driving, 3
Dyblie's rope-tightener, 30
Dynamo-driving, 38

Early use of ropes for driving, 2, 64
Effect of centrifugal force, 45
Effect of tar on ropes, 91
Effect of wedging in groove, 162
Efficiency in any given case, 73
Elasticity of ropes, 4, 162
Elastic slip of ropes, 174
Engines, friction in, 6, 142
English rope system, 4
Executed rope transmissions, 96
Experiments on friction, 141
Examples of large rope-wheels, 205 to 221

Fairbanks, Morse & Co., split pulley, 193
Fibrous ropes, 75
Fibre, cotton, 78
Fibres of manilla, 82
Flax ropes, 77
Fly-wheels, heavy, 5
Frictional grip, 167
Friction and stress moduli, 118
Friction-clutch, 33, 42
Friction, coil, 44, 48
Friction, coefficient of, 112
Friction loss, 141, 158
Friction of engines, 6
Friction of shafting, 6, 68, 145

Gear wheels, 1, 5, 8
Graphite on ropes, 89, 91
Gregg, multiple sheaves, 36, 38
Groove, angle, 164
Groove, shape of, 186
Groove, surface of, 183
Groove, wedging of rope in, 162
Guide-pulleys, 189

Harmonic vibration, 105
Heavy fly-wheels, 5
Hemp ropes, 2, 77, 90
Henthorn, J. F., friction in mills, 7
Hick, Hargreaves & Co., built-up pulley, 208
Hirn, C. F., transmission of power, 68
Hoadley Bros., Power-house, Chicago City Ry. Co., 52
Horse-power of ropes, 111 to 121
Hubs of pulleys, 191, 203
Hunt, C. W., form of splice, 20

Idlers, 34, 36
Inclined transmissions, 139
Influence of belt pull, 157
Influence of centrifugal force, 111, 119, 163
Introduction of rope-driving, 2

Jack-shaft, use of, for dynamo-drives, 38, 42

Jaw clutch, 42
Joint for braided ropes, 23

Kircaldy, tests on ropes, 80

Lambeth ropes, 80, 88
Lanett Mills, 12
Large rope-wheels, 205 to 221
Laxey, large overshot wheel at, 71
Least diameter of pulleys, 103, 161, 179
Leather belts, 1
Length of manilla fibre, 86
Life of ropes, 107
Limit of length in shafting, 69
Linseed-oil on ropes, 89
Link-belt Co., Western Electric Co.'s Plant, 25
Link-belt Co., Virginia Hotel Plant, 42
Liverpool Overhead Railway, 40
Lockwood & Greene, Lanett Mills, 12
Lockwood & Greene, Naumkeag Mills, 12
Long-distance transmission, 62
Loss due to bending, 159
Loss due to friction, 141, 158
Loss due to winder-pulleys, 50
Losses in ropes, 141, 159
Lubrication of ropes, 79, 88

Manilla fibre, 82, 86
Manilla ropes, 77
Marlin-spike, improved form of, 23
Method of joining arms and rim, 204
Miller, T. S., plant of Western Electric Co., 25
" " varying angle of groove, 168
" " use of small ropes, 100
" " use of loose idler, 36
Multiple idle sheaves, 36, 38
Multiple-rope system, 4, 17
Musgrave & Sons, Atlas Mills, 12
" " " , Nevsky Mills, 16
" " " , rope-pulley, 185
" " " , life of cotton ropes, 107

Naumkeag Mills, 12
Nevsky Mills, 16
Normal working load, 98, 100

Outdoor transmission, 50, 66
Overshot wheel at Laxey, 71

Pine tar as lubricant, 89
Pitch diameter of pulley, 181
Power absorbed by friction in shafting, 152
Power absorbed by ropes and gears, 6
Power transmitted by shafting, 153
Pulleys, diameter of, 38, 103, 161, 177, 181
Pulley, tightener, 48
Pulleys, supporting, 64
Pulley, winder, 47 to 54, 66
Pulleys, wood, 56, 164
" , with double arms, 193
" , with steel arms, 185
" , very wide, 203

Rawhide ropes, 25, 75
Relative cost of ropes, 122 to 127
Relative wear of ropes, 122
Rim sections, 166, 187
Ropes, cotton, 77, 81
" , cost of, 101, 122
" , early use of, 2
" , elasticity of, 162
" , fibrous, 75
" , flax, 77
" , hemp, 2, 77, 90
" , horse-power of, 121
" , Lambeth, 80, 88
" , life of, 107
" , lubrication of, 79
" , manilla, 77
" , rawhide, 25, 75
" , round leather, 76
" , shrinkage of, 31
" , speed of, 103
" , splicing of, 18

Ropes, steel and leather, 76
" , stevedore, 90
" , strength of, 79, 91
" , square leather, 76
" , wear of, 101, 104, 122
" , weight of, 109
" , wire, 75
Rope wells, 10
Rotation of ropes, 174

Sag of ropes, 131, 170
Sections of arms, 199
" " rim. 187
Section of rope, 92
Semicircular grooves, 164
Sewall, M. W., built-up pulleys, 217
Shafting, for long-distance transmissions, 68
" friction of, 145
" limit of length, 69
" loss due to friction, 68
" power transmitted, 153
Shafts, jack, use of, for dynamo-drives, 38, 42
" at an acute angle, 35
" at right angles, 36
Shrinkage of ropes, 31
Side lead of ropes, 35
Size of pulley, 38
" " ropes in use, 100, 107
Slack-side tension, 111
Small ropes, 24
Speed of ropes, 103
Split hubs, 191
Splicing of ropes, 18
Stevedore, transmission rope, 90
Stress in rim-bolts, 196
" " ropes, 67
Strength of ropes, 79, 91
Supporting pulleys, 64
Surface of groove, 183

Taper of arms, 203

INDEX.

Tallow on ropes, 89
Tar, effect of, on ropes, 91
Telodynamic transmissions, 63
Temporary installations, 52
Tension carriage, 27
" weight, 28, 32, 140
" in ropes, 111 to 118, 132
Tests on ropes, 80
Tightener, 29, 30
" pulley, 48
Transmissions at an angle, 34
" outdoor, 50, 66
" of power to a distance, 62
" telodynamic, 63
Turbines, Victor, 56
Twist in ropes, 86, 88

Use of ropes with portable tools, 62
Use of water-wheels, 56, 66, 71

Van Vleck, built-up pulley, 214
Vibration in ropes, 105
Victor turbine, 56

Walker, differential pulley, 173
Walker Mfg. Co., built-up pulley, 210, 220
Watertown tests on ropes, 80
Water-wheels and rope-driving, 56, 66
Wear of ropes, 83, 101, 104, 122
Wear due to side lead, 35
Weakening effect due to twisting, 86
Webber, Samuel, early use of ropes, 2
" " friction of shafting, 8
Wedging action, 162, 169
Weight for tension-carriage, 139
Weight of ropes, 109
Wells, rope, 10
Western Electric Co.'s plant, 25
Wetherill & Co., built-up pulley, 221
Wide pulleys, 203
Willamette Mills, 66
Winder-pulley, 47 to 54, 66

Wire ropes, 75
Wood pulleys, 56, 164, 186
Wood-filled pulley rim, 186
Working strength of ropes, 95, 98
Wound system of rope-driving, 25
Wren's "Instrument for drawing up great weights," 45

www.ingramcontent.com/pod-product-compliance
Lightning Source LLC
Chambersburg PA
CBHW031742230426
43669CB00007B/445